The HISTORY of SCIENCE

in BITE-SIZED CHUNKS

The HISTORY of SCIENCE
in BITE-SIZED CHUNKS

NICOLA CHALTON &
MEREDITH MACARDLE

Michael O'Mara Books Limited

This paperback edition first published in 2019
First published as *The Great Scientists in Bite-sized Chunks*
in Great Britain in 2015 by
Michael O'Mara Books Limited
9 Lion Yard
Tremadoc Road
London SW4 7NQ

Copyright © Basement Press 2015, 2019

All rights reserved. You may not copy, store, distribute, transmit, reproduce or otherwise make available this publication (or any part of it) in any form, or by any means (electronic, digital, optical, mechanical, photocopying, recording or otherwise), without the prior written permission of the publisher. Any person who does any unauthorized act in relation to this publication may be liable to criminal prosecution and civil claims for damages.

A CIP catalogue record for this book is available
from the British Library.

Papers used by Michael O'Mara Books Limited are natural, recyclable products made from wood grown in sustainable forests. The manufacturing processes conform to the environmental regulations of the country of origin.

ISBN: 978-1-78929-071-4 in paperback print format
ISBN: 978-1-78929-177-3 in ebook format

1 3 5 7 9 10 8 6 4 2

Designed and typeset by Basement Press (www.basementpress.com)
Illustrations by Pascal Thivillon

Printed and bound by CPI Group (UK) Ltd, Croydon CR0 4YY

www.mombooks.com

To Felix, who wants to be a scientist one day.

Contents

Introduction	9
CHAPTER 1 Astronomy and Cosmology: A Scientific View of the Universe	13
CHAPTER 2 Mathematics: The Science of Numbers	53
CHAPTER 3 Physics: What Things are Made Of	81
CHAPTER 4 Chemistry: Discovering Elements and Compounds	111
CHAPTER 5 Biology: Characteristics of Life on Earth	137
CHAPTER 6 The Human Being and Medicine	165
CHAPTER 7 Geology and Meteorology	195
Index	219

Introduction

SCIENCE TODAY HAS two meanings: the investigation of the world around us, and the way in which such investigations are carried out – the scientific method. Different branches of science explore literally everything in the universe, from its origins to its tiniest particles, from the human body to rocks and minerals, and from the power of lightning to invisible forces such as X-rays, radioactivity and gravity.

While our earliest human ancestors probably looked up at the night sky and wondered how the world came about, or collected the first medicinal plants, the scientific method itself is relatively new. Many early investigators just proposed their own particular hypothesis about something, and did not think of testing that theory through carefully conducted experiments that could be repeated again and again to reach the same results. Today it would be unthinkable for a scientist not to offer thoroughly tested evidence for a new theory. In some disciplines, such as astronomy, it is not always possible to carry out experiments, but predictions of events and observations can be used to verify, or reject, a hypothesis.

Early proponents of some form of empirical scientific method include ancient Greek philosopher–scientists, the Arab optical specialist Ibn al-Haythem, the medieval English monk Roger Bacon and the Italian astronomer Galileo Galilei. But the major change in scientific attitudes came in the seventeenth century with the approach by one of science's great minds, Isaac Newton. He proposed 'rules of reasoning' that encompassed propositions and experiments, and very soon all investigators into nature were adopting his method.

In most scientific disciplines there is an understanding that a proven hypothesis might be accepted as a scientific 'truth', but only until a new theory might disprove it and offer a new paradigm. In this way science develops and expands as new ideas displace old ones. The exception is mathematics, where once a theorem is shown to be true, it is permanently so. It can never be disproven. In fact mathematics, though scientific in the broader sense of involving systematic and formulated knowledge, differs quite considerably from the natural sciences, which investigate the physical universe. Where the natural sciences gather empirical evidence to devise and refine descriptions or models of aspects of the physical universe, mathematics gathers proofs for necessary truths. However, mathematics provides the language in which the natural sciences aspire to describe and analyse the universe, and in this respect it is strongly linked to the sciences.

Science is often coupled with technology too, for many scientific discoveries and advances lead to technological changes: the invention of the light bulb, credited to Thomas

Edison, relied upon centuries of scientific exploration of electricity; exploration of space has given us calendars and advanced ceramic technology as used in spacecraft, among many other benefits. From medical technology to the computers and smartphones that we could not live without, science has impacted on everyday life in every way.

Of course, science would not exist without people who are passionate about finding out how the world works. This book reveals some of the great scientists throughout history who have shaped our understanding of the universe.

CHAPTER 1

Astronomy and Cosmology: A Scientific View of the Universe

SINCE ANCIENT TIMES, humans have tried to make sense of the universe by observing objects beyond our world – the sun, moon, stars and planets. Babylonian and Egyptian civilizations, realizing that astronomical events are repeated and have cycles, charted star positions and predicted celestial events such as eclipses, comets and the motions of the moon and the brightest stars. Their records formed the basis for timekeeping and navigation.

Adopting centuries of observations before them, the ancient Greeks named groups of stars, or constellations, after mythological figures like Orion, the hunter, and Gemini, the twins Castor and Pollux. The forty-eight Western constellations listed by Ptolemy in the first century are among the eighty-eight constellations used to navigate the night skies today. Likewise, the Romans gave us the names for some of our planets: Mercury, Venus, Mars, Jupiter and Saturn. Reflecting the sun, these were seen as bright 'stars' in the sky.

The invention of the optical telescope in the seventeenth century changed forever the idea of an earth-centred universe.

It was soon clear that the universe was much larger than ever imagined. Searching deeper into space, astronomers found more planets of our solar system (Uranus and Neptune), minor planets or asteroids, satellites (moons), dwarf planets (like Pluto), gas clouds, cosmic dust and whole new galaxies.

Today's astronomical instruments include satellite-borne telescopes that detect radiation from faraway cosmic objects, and space probes that can bring back information from other planets. Armed with these tools, astronomers are discovering more about the particles and forces that make up the universe, the processes by which stars, planets and galaxies evolve, and how the universe began. They have also discovered a large part of the universe that cannot be seen with any type of telescope. This 'dark matter' is proving to be one of astronomy's greatest mysteries.

Early Star Catalogues: Gan De

Chinese astronomer Gan De (*c.* 400 to *c.* 340 BCE) and his contemporary Shi Shen are believed to be the first named astronomers in history to compile a list of stars, or star catalogue. Gan De lived during the turbulent Warring States period of ancient China, when the regular twelve-year passage across the skies of the bright, visible light of Jupiter, the largest planet in our solar system, was used to count years, so it was the focus of concentrated observations and predictions. Without telescopes, Gan De and his colleagues had to rely on the naked eye, but they made acute calculations to guide them as to the best times to make celestial observations.

In the night sky above mainland China, Gan De saw and catalogued more than a thousand stars, and he recognized at least a hundred Chinese constellations. His star catalogue was more comprehensive than the first-known Western star catalogue, drawn up 200 years later by the Greek astronomer Hipparchus, who listed about 800 stars.

Gan De's observation of what was almost certainly one of Jupiter's four large moons was the first-known record in the world of seeing a satellite of Jupiter – long before Galileo Galilei officially 'discovered' the satellites in 1610 using his newly developed telescope.

Shi Shen and Gan De were among the first astronomers to approach an accurate measurement of a year, at 365¼ days. In 46 BCE the Greek astronomer Sosigenes of Alexandria would be employed by Julius Caesar to realign the Roman calendar to this more accurate measurement. The resulting Julian calendar remained in use across Europe and Northern Africa until 1582 and the introduction of the Gregorian calendar that is still in operation today.

Geocentric View of the Cosmos: Aristotle

In the fourth century BCE, while the ancient Chinese states battled for supremacy, classical Greek culture was spreading to numerous colonies around the eastern Mediterranean, laying the foundation stone that would underpin Western thought into the modern era.

The Greeks felt they were at the centre of the cosmos, and the night skies added to their belief. Stars appeared to rise and then set, as if on a journey around the earth. (The illusion is a result of the earth spinning on its axis: stars appear to move westwards across the sky simply because the earth rotates eastwards.)

They identified 'wandering stars', whose positions move in relation to the 'fixed stars' that twinkle in the background. The wanderers were the sun and moon and five then-known planets of our solar system: Mercury, Venus, Mars, Jupiter and Saturn. The Greeks concluded that the cosmos, or universe, consisted of the earth, a perfect sphere (not flat as archaic cultures had thought) that was stationary at the centre of everything, with heavenly bodies – the sun and visible planets – orbiting in uniform motions and perfect circles around the earth. The 'fixed stars' were located in the outer celestial sphere – astronomers didn't notice the actual movements of these faraway stars until the nineteenth century.

To this 'geocentric theory' the great natural philosopher and scientist Aristotle added his own ideas. The earth and heavens, he theorized, were made up of five elements: four earthly elements (earth, air, fire and water) and a fifth element, a material filling the heavens and arranged in concentric shells around the earth, called aether. Each concentric shell of aether contained one of the heavenly bodies, orbiting around the earth at a uniform pace and in a perfect circle. In the outermost shell the stars were all fixed. The earthly elements came into being, decayed and died, but the heavens were perfect and unchanging.

Aristotle's cosmological ideas were accepted in the Arab world and reintroduced into Christian Europe during the Middle Ages.

ASTRONOMY AND COSMOLOGY

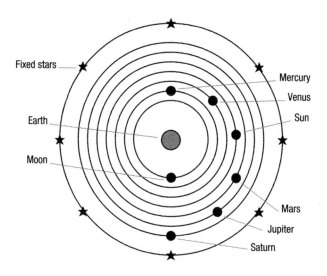

The geocentric cosmological model was the prevailing view in ancient Greece.

Aristotle (384–322 BCE)

Aristotle was a giant of the classical Greek intellectual world and his ideas had a lasting influence in the West. Born to a Macedonian medical family, he was one of the stars of Plato's school in Athens.

He left Athens possibly because he was not appointed head of the Academy after Plato's death, and perhaps also because Philip of Macedon's expansionist wars had made Macedonians unpopular. But he returned to the city in 335/34 BCE after Alexander the Great – Philip's son and Aristotle's pupil – had conquered all of Greece.

While running his own school in Athens, the Lyceum, Aristotle continued extensive studies into almost every subject then defined. His method of teaching and debate was to walk around discussing topics, which is why Aristotelians are often called Peripatetics.

Following Alexander's death, ill feeling towards Macedonians flared up again and Aristotle fled, supposedly stating in a reference to the execution of the philosopher Socrates seventy years earlier: 'I will not allow the Athenians to sin twice against philosophy.'

Precession of the Equinoxes: Hipparchus

Classical Greek culture flowed eastwards in the wake of Alexander the Great's conquests, inspiring scholars like Hipparchus (*c.* 190 to *c.* 120 BCE) of Nicaea (in what is now Turkey).

While compiling a star catalogue, Hipparchus noticed that the positions of stars did not match earlier records: there was an unexpected systematic shift. He concluded that the earth itself had moved, rather than the stars. He had detected the earth's 'wobble' as it rotates around its axis – imagine the slow wobble of a spinning top, with its axis tracing a circular path. One such circuit caused by the earth's wobble takes about 26,000 years – a figure calculated very accurately by Hipparchus.

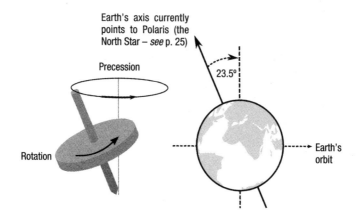

The precession of the equinoxes – the earth tilts on its axis at an angle of 23.5 degrees, and wobbles like a top as it rotates, but only slowly: one wobble (or 'circle of precession') every 26,000 years. The wobble affects the equinoxes, or the timing of the seasons.

He named the wobble the precession of the equinoxes because it causes the equinoxes (the two dates each year, in March and September, when day and night are of equal length) to occur slightly earlier than expected with reference to the 'fixed stars'.

Over time this discrepancy caused the seasons to occur at different times in the ancient calendar systems. These were based on the solar measurement for a year (the 'sidereal year'), which is the time it takes for the sun to revolve from a position in the sky marked by a fixed star to the same position again, as viewed from the earth (or, as we now know, the time it takes for the earth to orbit once around the sun). Hipparchus solved the problem by inventing a new measurement for a year, the 'tropical year', or the time it takes for the sun's apparent revolution from an equinox to the same equinox again. About twenty minutes shorter than the sidereal year, the tropical year is the basis for our modern Gregorian calendar. It ensures that the seasons occur in the same calendar months each year.

Hipparchus used Babylonian data to calculate the lengths of the sidereal and tropical years with great accuracy: indeed, far more accurately than Ptolemy, who came about 250 years later, showing just how far Hipparchus was ahead of his time.

A Mathematical Cosmos: Ptolemy

Ptolemy, born towards the end of the first century, and the last of the great ancient Greek astronomers, also adopted the geocentric view of the earth at the centre of the cosmos. His contribution was to create the first model of the universe that would explain and predict the movements of the sun and planets in mathematical terms. His model appeared to answer a question that had puzzled the Greeks for some 1,400 years: why, if a planet was orbiting around the earth at the centre of the universe, did it sometimes appear to move backwards with respect to the positions of the 'fixed stars' behind it?

Despite Ptolemy's fundamental beliefs, in order to explain mathematically the movements of heavenly bodies he had to violate his own rules by assuming that the earth was not at the exact centre of the planetary orbits. Pragmatically, he and his followers accepted this displacement, known as the 'eccentric', as just a minor blip in the essential geocentric theory.

Ptolemy used a combination of three geometric constructs. The first, the eccentric, was not new, nor was his second construct, the epicycle. This proposed that planets do not simply orbit the earth in large circles, but instead move around small circles, or epicycles, which in turn revolve around the circumference of a larger circle (the deferent) focused (eccentrically) on the earth. Progress along the epicycle explained why planets sometimes appear to move backwards, or in 'retrograde motion' (*see* diagram, next page).

His third construction – the equant – was revolutionary, and Ptolemy invented it to explain why the planets sometimes seemed to move faster or slower, rather than uniformly, as viewed from earth. He suggested that the epicycle's centre of motion on the circumference of its larger circle (the deferent) is not aligned with either the earth or the eccentric centre of this larger circle, but with a third point, the equant, which is situated opposite the earth and at the same distance from the centre of the larger circle as is the earth. It is only from the point of view of the equant that the planet appears to be in uniform motion.

These three mathematical constructions – epicycle, eccentric and equant – were complex and unsatisfactory to purists, but they seemed to explain some puzzling aspects of astronomy, such as the retrograde motion of planets and why planets appear brighter, and therefore closer to earth, at different times. Together they allowed predictions of planetary positions that approximated those of a modern heliocentric view of the universe, in which the planets orbit the sun in elliptical paths.

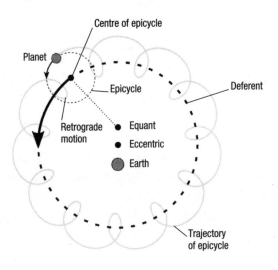

Ptolemy's geocentric model positioned earth not at the exact centre of the planetary orbits and seemed to explain the retrograde (backwards) movement of planets.

Ptolemy's geocentric model was first followed in the Middle East and then in Western Europe. It aligned with religious belief, and scholars who dared dispute it faced a death sentence from the rigid and repressive Catholic Church. But by 1008, Arab astronomers were questioning Ptolemy's data and his ideas, and centuries later it was clear that at least some of his recorded observations were falsified to match his theories.

Ptolemy (*c.* 83 to *c.* 161 CE)

Ptolemy lived in Egypt when it was a province of the Roman Empire. Although his first name, Claudius, is Roman, his Latin surname, Ptolemaeus, suggests Greek heritage and he used Greek for his writings. He made his observations of the heavens from the town of Alexandria, whose magnificent library was a magnet for ancient scholars of all disciplines.

Until Ptolemy began writing, there had been a gap of 200 years in Greek astronomy since the days of Hipparchus, and it is only thanks to Ptolemy that we know about Hipparchus' work. Ptolemy was a great synthesizer, and he acknowledged using earlier theories in his explanation of how the universe works.

Astronomical Records in the Arab World: al-Battani

Ptolemy's predictions of planetary positions were surpassed in early medieval times by a brilliant Arab astronomer and mathematician, al-Battani (c. 858 to 929).

Descended from a well-known instrument maker and astronomer, al-Battani lived at a time when the Muslim empires encouraged learning and kept alive the science and philosophy of ancient Greece and Rome. At the crossroads between East and West, Muslim scholars also received ideas from the Asian civilizations of China and India, and incorporated them, with their own discoveries, into a body of knowledge that was later passed on to Europe.

Al-Battani produced a set of elaborate astronomical tables that recorded positions of the sun, moon and planets and could be used to anticipate their future positions. His 'Sabian Tables' were the most accurate available at the time and they would influence the Latin world.

Rather than employ geometrical methods, as other astronomers had done before him, al-Battani used trigonometry for astronomical calculations. He made the amazingly accurate determination that our solar year is 365 days, 5 hours, 46 minutes and 24 seconds long, only a couple of minutes off today's reckoning of 365 days, 5 hours, 48 minutes, 45 seconds. And he made a discovery that had eluded Ptolemy: the earth's distance from the sun, and the moon's distance from the earth, vary throughout the year. As a result he correctly predicted annular solar eclipses, in which the moon covers the sun's centre, leaving a 'ring of fire' around the moon.

Al-Battani was so well regarded that the groundbreaking mathematician and astronomer Nicolaus Copernicus would acknowledge his work 600 years later.

Pole Star Navigation: Shen Kuo

In the eleventh century, sea navigators relied upon landmarks and observations of celestial bodies, including the northern pole star, or 'North Star'.

The North Star lies roughly in line with the earth's axis and directly over your head if you stand at the North Pole. As the earth spins on its axis, an observer in the earth's northern hemisphere has the impression of stars rotating around earth – except for the North Star, which stays put, making it an excellent navigational pointer for the geographic North Pole. It can also be used to determine latitudinal position (north–south coordinates) by measuring its height above the horizon.

Since late antiquity Polaris has taken the role of the northern pole star, but due to the precession of the equinoxes, the earth's slow-motion wobble as it turns on its axis (*see* p. 18), the North Pole will point to a different pole star, Gamma Cephei, in about 3000 CE; and in about 15,000 CE the role will pass to Vega. The pole star will be Polaris again in the distant future.

Chinese polymath and government official Shen Kuo (1031–95 CE), with his colleague Wei Pu (active 1075 CE), measured the position of the pole star every night for five years. Later he recorded the invention of a magnetic needle compass in China, used by sailors across Europe and the Middle East. He was the first person to discover that magnetic needles point to the magnetic north and south, not to geographic, or true, north and south.

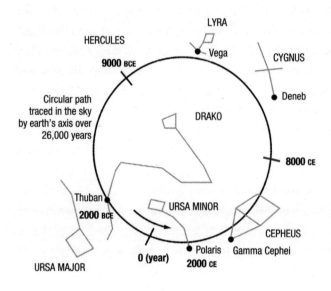

Polaris will not be the North Star forever because of the earth's wobble (the precession of the equinoxes).

Arabic Tables and Astrolabes: Azarquiel

Azarquiel (1028–1100), also known as al-Zarqali, was born in the Spanish Muslim town of Toledo, an important centre of learning that suffered constant attacks by Spanish

Christians. He made delicate scientific instruments for a living until his clients encouraged him to get an education in mathematics and astronomy. Later he compiled the Toledan Tables, widely accepted as the most accurate astronomical tables so far produced, and by the twelfth century used throughout Europe.

The Toledan Tables helped astronomers to predict the movements of the sun, moon and planets relative to the 'fixed' stars, and solar and lunar eclipses many years ahead. They were adapted for different locations in the Christian West and formed the basis for the Alfonsine Tables (*c.* 1252 to 1270), which would endure in Europe far into the sixteenth century.

Azarquiel made another major contribution when he developed a new type of astrolabe. Hipparchus had invented a predecessor in about 150 BCE, but Azarquiel's instrument could be used at any latitude to measure the altitude of the sun, moon and stars, and determine latitude. In the medieval Arab world astrolabes were important for scheduling prayers; eventually they would be developed for navigation at sea.

Celestial Navigation and the Age of Exploration: Abraham Zacuto

Fifteenth-century Jewish scientist and rabbi Abraham Zacuto was born in Spain at a time when most European sailors followed well-known routes hugging the coastline. Zacuto would change all that with his navigational equipment that allowed European explorers to sail across the oceans to the Americas and East Indies.

One of Zacuto's great achievements was to develop solar tables for navigation during daytime (the pole star was used at night). With an astrolabe adapted for marine use and these tables, a sailor could determine the latitude of a ship based on the sun's altitude (which varies at different times of the year) as measured from the ship. The metal astrolabe was held vertically, with the zero mark on the circular disc aligned with the horizon, and the moveable sight bar aimed at the sun; the altitude was read off the degree scale. Also, comparing the sun's altitude at a sea location with the altitude recorded at a point of departure (say, Lisbon), navigators could calculate their distance north or south from Lisbon.

These techniques were used to develop sea charts and proved invaluable for sailors venturing into unknown waters, including explorers Bartolomeu Dias, Vasco da Gama and Christopher Columbus.

Zacuto published an almanac with tables of celestial events that once famously saved the life of Christopher Columbus. During his fourth voyage to the New World, Columbus and his crew were at risk of being killed by a group of natives but Columbus knew from Zacuto's almanac that a total lunar eclipse was due on 29 February 1504, so he used it to his advantage, saying that the disappearance of the moon would show that his god was angry with them. When the moon reappeared he announced it was a sign that the natives had been pardoned, which changed their attitude pretty quickly!

Two hundred years later, the more accurate sextant would replace the astrolabe and become the standard instrument for celestial navigation, though mariners would have to wait until the chronometer was invented in the eighteenth century before they could measure longitude and properly fix their position in the open seas.

Abraham Zacuto (*c.* 1452 to *c.* 1515)

The long-established Jewish communities in the Iberian peninsula benefited from contact with Arab cultures, producing many great scholars. Zacuto was one of them, a Renaissance man with wide-ranging interests. He encouraged his friend, the explorer Christopher Columbus, to persevere with his dream of sailing to Asia.

When, in 1492, the Spanish monarchs Ferdinand and Isabella demanded that Jews either convert to Christianity or leave Spain, Zacuto left for Portugal and settled in Lisbon. He soon gained a position as royal astronomer and historian. Consulted by King Manuel and by the sailor Vasco da Gama, Zacuto agreed that a voyage of exploration to the East would be feasible. That same year King Manuel issued an ultimatum to Portuguese Jews to convert or leave. Zacuto and his son Samuel were among the few to escape in time, but on their journey to sanctuary in North Africa they were twice captured by pirates and held to ransom.

Zacuto eventually landed in Tunis, but the ever-present fear of Spanish invasion forced him to move on, and he wandered around North Africa before settling in Turkey.

Solving the Longitudinal Problem: John Harrison

The advance that marine navigators had been waiting for came in the 1770s when a self-educated English clockmaker, John Harrison (1693–1776), solved the 'longitudinal problem' by inventing the seagoing chronometer.

Up until then, sailors had struggled to determine their longitudinal position (east–west coordinate). Italian explorer Amerigo Vespucci (1454–1512) complained:

> As to longitude, I declare that I found so much difficulty in determining it that I was put to great pains to ascertain the east–west distance I had covered. The final result of my labours was that I found nothing better to do than to watch for and take observations at night of the conjunction of one planet with another, and especially of the conjunction of the moon with the other planets, because the moon is swifter in her course than any other planet. I compared my observations with an almanac.

His method gave a rough estimate of longitude but could only be used when a specific astronomical event was predicted and it required knowledge of the precise time, which was difficult for sailors far from home.

Another practice involved comparing the local time of the ship's position at sea (by looking at the position of the sun) with the time at a known location, for example the ship's starting point, recorded on a clock on board, to estimate how far eastwards or westwards the ship had sailed. How this works becomes clear when you consider that lines of longitude are drawn so that every 15 degrees of longitude (travelling eastwards or westwards) correspond to either one hour ahead or back in local time. Again, the problem was to know the precise time.

Harrison's marine chronometer, or portable 'sea watch', provided a solution. Far more accurate than the best watches available, it could withstand weather variations at sea and the yawing action of a rolling ship. British explorer Captain James Cook was full of praise for the instrument when he circumnavigated the globe in 1772–5. The model he used can now be seen in the National Maritime Museum in London.

In 1884 the prime meridian (0 degrees longitude) at Greenwich in England was established, and from this date onwards every place on earth has been measured in terms of its distance east or west from this line. Modern ships use satellite navigation systems to record location precisely, but often they will carry a chronometer on board in case of emergencies.

Modern Astronomy Begins: Nicolaus Copernicus

The publication of Copernicus' heliocentric theory in 1543 brought the first serious challenge to the geocentric cosmological system. Johann Wolfgang von Goethe would later comment: 'Of all discoveries and opinions, none may have exerted a greater effect on the human spirit than the doctrine of Copernicus. The world had scarcely become known as round and complete in itself when it was asked to waive the tremendous privilege of being the centre of the universe.'

Ptolemy's geocentric model, given credibility by certain passages of Scripture, had prevailed in Europe for the last 1,500 years. It accorded with the appearance of the skies to any casual onlooker, and by placing man at the centre of things it appealed to human nature. But Copernicus saw the logic in heliocentrism: 'In the centre rests the sun. For who would place this lamp of a very beautiful temple in another or better place than this where from it can illuminate everything at the same time?'

Copernicus' astronomical system had the great advantage of simplicity. It did not require the complex sets of geometrical techniques to explain the motion of the planets, characteristic of Ptolemaic astronomy, because it recognized that their apparent backward movement was only perceived, not real, and was due to the motion of the earth. It put the sun at the centre of things and revolving around it, in order, were Mercury, Venus, the earth and the moon, Mars, Jupiter and Saturn, and beyond was a large sphere of fixed stars. The earth rotated around its axis daily, the moon revolved around the earth once a month, and the earth, tilted on its axis, revolved around the sun annually.

Heliocentrism would cause a public outcry, upset the Church and signal the beginning of the Scientific Revolution.

Nicolaus Copernicus (1473–1543)

Born to a wealthy Polish family, Copernicus would have learned about Ptolemy's theory at the University of Krakow, where he studied astronomy. He was appointed canon at the cathedral of Frauenberg in 1501 and his clerical position allowed him time to study 'astrological medicine' – physicians in medieval Europe made use of astrology in the belief that the stars could influence the course of human affairs.

Copernicus was apparently unaffected by the turbulent years of the Reformation, spending many hours in solitary study making celestial observations from a turret in the fortifications of the town. He had no instruments to help him as the telescope would not be invented for another hundred years.

Around 1514, Copernicus circulated among a few friends an early stage in the development of his heliocentric theory. A detailed exposition was printed on his deathbed. The Lutheran minister overseeing the printing inserted an anonymous preface presenting the theory, contrary to Copernicus' opinion, as a practical, mathematical device for charting the movements of the planets, and not as a truth about the world. It wasn't until the early seventeenth century that the Church was provoked to object to heliocentrism.

Championing Heliocentrism: Johannes Kepler

After Copernicus, the first astronomer openly to support a heliocentric theory of the universe was German astronomer and mathematical genius Johannes Kepler.

A committed Christian, Kepler believed that God had used a geometrical plan to build the cosmos; he thought he would be drawn closer to his maker if he could understand God's plan.

Using Euclid's geometry, he developed a model for the orbital paths of each of the known planets and found the sun to be the point around which all the planets circled. He concluded that the sun was the centre and mover of the other planets. His model mirrored his view of the spiritual universe: God the Father, like the large and powerful sun, at the centre of creation. It also matched heliocentrism.

He then spent years struggling to understand the erratic orbit of Mars (he happened to have access to data on the planet), calling it his 'War against Mars', until he realized that his underlying assumption was wrong: 'It seemed as if I woke from sleep and saw a new light break on me.' The orbits of each of the known planets around the sun could not be perfect circles, as in Copernicus' model, but were oval-shaped ellipses with the sun as one focus (this became Kepler's first 'law'). When a planet is closest to the sun, it is moving fastest, and when it is farthest from the sun, it is moving slowest. Nonetheless, an imaginary line from the centre of the sun to the centre of a planet sweeps out equal areas of space during equal intervals of time. This became his important second 'law' of planetary motion and it can be used to determine how fast a planet will be moving at any point in its orbit.

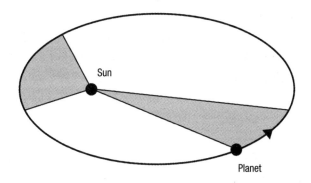

Kepler's Second Law: The line joining the planet to the sun sweeps out equal areas in equal intervals of time.

His third 'law' used geometry and knowledge of a planet's orbital period to calculate its distance from the sun.

Kepler's laws of planetary motion helped to overthrow the belief in heavenly circles that had dominated cosmology for more than 2,000 years. Eighty years later Isaac Newton would provide mathematical explanations for Kepler's theories and use them to construct his work on universal gravitation (*see* p. 40).

Kepler laid the foundation for much of our modern understanding of the workings of the solar system and gave us practical laws – we can use them, for example, to calculate the orbits of artificial satellites (a word he coined) and spacecraft.

Although a useful model of the solar system, astronomers would come to realize that the heliocentric theory is not strictly true, as the sun is not the centre of the universe, but just one of innumerable stars.

Johannes Kepler (1571–1630)

Kepler lost his father, a mercenary soldier, when he was five, and his interest in astronomy began a year later when his mother took him to a hilltop to see a comet.

A life-long Lutheran, he intended to become a minister, but as was customary he took courses in other subjects and so he learned about, and eventually supported, the new heliocentrism. He taught mathematics and astronomy at the Protestant school in Graz (in modern Austria) but Europe's growing religious strife intervened – as it was to do several times in his life. Protestants were banished from Graz, and Kepler and his family took refuge in Prague, where he helped the Danish astronomer Tycho Brahe (1546–1601) to prepare a new set of astronomical tables. After Brahe's sudden death in 1601 Kepler was appointed his successor as imperial mathematician to the Holy Roman Emperor Rudolf II and entrusted to finish the tables.

Kepler experienced the full gamut of seventeenth-century religious pressure when, in 1620, his mother

Katharina was imprisoned, accused of witchcraft and threatened with torture. Kepler assisted with her case and she was released, but only after a protracted legal battle.

When Prague turned against Protestants, Kepler had to move to Linz (in modern Austria), and he moved again in 1626 when Catholic forces besieged Linz during the Thirty Years' War. His work disrupted by religious conflict, and worn down by his unsettled existence, Kepler succumbed to a fever and died in Regensburg (in south-east Germany). His grave was lost but his epitaph survives:

> I used to measure the skies,
> now I shall measure the shadows of the earth.
> Although my soul was from heaven,
> the shadow of my body lies here.

The Telescope that Revolutionized Astronomy: Galileo Galilei

Galileo is best known for building the first telescope powerful enough through which to observe the solar system in some detail.

His improvements to the design of the telescope in 1609 meant that he was the first person to turn an effective magnifying instrument upon the heavens, and the first person to report seeing the craters and mountains of the moon. He made other original observations in 1610 that brought new knowledge of the solar system: the four largest moons orbiting Jupiter, showing that at least some heavenly bodies did not go around the earth; the phases of Venus, suggesting that it orbited the sun; and a vast number of stars, indicating that the universe was far larger than previously imagined.

All in all, Galileo concluded that the Church was wrong to hold that the sun and other planets orbited the earth. He wrote in a letter in 1615: 'With regard to the movement of the sun and earth, the inspired Scriptures must obviously adapt themselves to the understanding of the people.'

Sometimes known as 'the father of modern science', Galileo used a quantitative experimental method that became the standard scientific approach. It involves carefully controlled and repeatable experiments to test a particular hypothesis (idea) about the natural world, then expressing the results mathematically, and then, depending on how well the test results match the predictions made from the hypothesis, either refining the original hypothesis or concluding that it is false. The process is ongoing and the aim is to reach a theory that is well supported by the evidence.

Galileo Galilei (1564–1642)

Galileo was born in Pisa, Italy, to a noble but impoverished family. His father hoped he would become a well-paid doctor and sent him to university. But, bored with everything except mathematical problems and natural philosophy, Galileo left without a degree.

Although he gained a reputation as a mathematician, he became desperately poor, as his father had feared. He turned to inventing and his fortunes improved dramatically with his creation of a telescope modelled on a Dutch invention he had never seen. He refined the instrument, eventually producing one that would enable him to make amazing astronomical discoveries, including evidence that the earth and planets revolve around the sun. His new fame brought him a lucrative role as court mathematician to Cosimo de Medici, the Grand Duke of Tuscany.

Galileo's support for the heliocentric system clashed with Church doctrine. In 1600 the Papal Inquisition had burnt the philosopher and cosmologist Giordano Bruno at the stake, so perhaps with this example in mind, Galileo recanted. He became a symbol of the tension between religion and scientific knowledge.

Rings of Saturn: Christiaan Huygens and Giovanni Cassini

Saturn, the second largest planet in our solar system, has been watched for millennia as a beautiful, bright yellow star in the night sky.

When Galileo first trained his telescope on this planet, in 1610, he thought he detected two moons on either side of it. Forty-five years later and with a more powerful telescope, Dutch astronomer Christiaan Huygens (1629–95) observed what he thought was a solid 'thin, flat ring' around the planet, and a moon (it was Titan, the largest of Saturn's satellites). Then in 1675 Italian astronomer Giovanni Cassini (1625–1712) found a gap between the rings (the 'Cassini division'), and four more moons.

Fast-forward to 2004 and the Cassini robotic spacecraft made the first ever orbit of Saturn and sent back pictures of the intricate bands of rock and ice that orbit around this gassy giant, constituting Saturn's distinctive rings. In trying to understand how and why the ring structure has developed, the ongoing mission hopes to gain insights into the origins and evolution of our solar system.

The mission also discovered water vapour and ice spewing from geysers at the south pole of Enceladus, another of Saturn's moons (it has more than sixty), and scientists now think that an underground ocean below the moon's icy crust is a likely explanation. A wet environment could, of course, be favourable to microbial life. The discovery has expanded our view of places in our solar system that might support life.

Cement of the Universe: Isaac Newton and Albert Einstein

British mathematician and physicist Isaac Newton (*see* p. 69) provided the first scientific explanation for how the universe is physically held together.

In 1684 the astronomer Edmond Halley (1656–1742) consulted Newton about planetary orbits. He was astounded to find that Newton had a complete scientific theory: gravity as a universal force holding together the structure of the universe.

Newton showed that the same force – gravity – acts both at close quarters and at vast distances: pulling an apple to earth and holding the planets in orbit around the sun. An object with more matter, or mass, exerts the greater force of attraction.

Newton published his work *Philosphiæ Naturalis Principia Mathematica*, often referred to as the *Principia*, in 1687. It covered gravity and the laws of motion and became the accepted scientific view of the universe.

Two hundred years later, Jewish physicist Albert Einstein (*see* p. 97) would advance Newtonian physics, transforming our understanding of space, time and gravity. In his general theory of relativity (1916) he explained that gravitation is not a force as Newton described, but a curved field caused by the presence of mass. Matter and energy warp the geometry of space, a bit like the way a heavy body sags a mattress, and this is the effect we call gravity. One consequence is that even light rays get bent by gravity, following a curved path around massive objects like the sun.

Einstein's theory was dramatically confirmed during the solar eclipse of 1919 when British astronomer Arthur Eddington (1882–1944) obtained evidence that the light from a distant star, positioned behind the sun as viewed from earth, had curved around the (blacked-out) sun to reach earth.

The Big Bang and the Origin of the Universe: Georges Lemaître and Edwin Hubble

A Belgian Jesuit priest and astronomer, Georges Lemaître (1894–1966), first came up with the idea of an expanding universe in 1927 and proposed what has developed into the 'Big Bang' theory of the origin of the universe.

Lemaître hypothesized that the expansion of the universe can be traced back to a single point in time (13.8 billion years ago is the modern estimate): the cataclysmic explosion of an extremely compact and dense 'Primaeval Atom', or 'Cosmic Egg'. His published findings were not widely read outside Belgium until 1931, when his solution was described as 'brilliant' by British astronomer Arthur Eddington, who had helped to translate it.

But it was Lemaître's better-known contemporary, American astronomer Edwin Hubble (1889–1953), who helped prove the theory of the expansion of the universe and the Big Bang model, earning him the title of founder of the science of cosmology.

Hubble started his professional career during exciting times. Henrietta Leavitt (1868–1921) had noticed that the Large and Small Magellanic Clouds (objects visible close to the edge of our Milky Way galaxy, now known to be dwarf galaxies) contained thousands of stars with variable brightness. Her observations led to the development of a method for measuring distances between stars that revolutionized our picture of the universe. Astronomers began to realize that the universe was much larger than previously thought. Albert Einstein's general theory of relativity then predicted the idea of a changing universe, expanding or contracting. Many, including Einstein himself, found this new view difficult to accept.

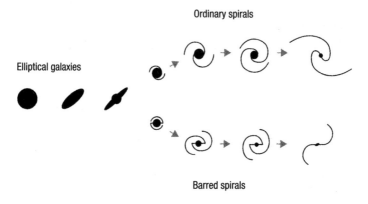

Hubble's 1936 classification of galaxies. Our galaxy, the Milky Way, appears as a dim glowing band across the night sky and is classed as a barred spiral galaxy. It consists of a flat rotating disc some 100,000 light years wide, containing gas, dust and roughly 100 billion stars. Our solar system is not at the centre of the galaxy but located on a minor spiral arm.

Hubble's contribution began with fuzzy patches of light, known as spiral nebulae, found throughout the night sky. Were these gas clouds located within our galaxy or groups of stars far beyond our galaxy? At the Mount Wilson Observatory in California, which housed the 254-cm (100-inch) Hooker telescope, the largest telescope in the world at the time, he focused his observations on a part of the sky called the Andromeda nebula and for the first time the images revealed faint stars. In 1923 he concluded that they were too distant to be part of our galaxy and so belonged to an entirely new galaxy (now known as the Andromeda galaxy), at least ten times further away than the farthest stars in the Milky Way. Further investigations led to the discovery of several other galaxies. Clearly we were part of a much larger universe than previously imagined. Hubble compared the galaxies and established a method for classifying them that is still used today.

Another important discovery came in 1929 when Hubble published his data on uniform universe expansion. By studying forty-six galaxies he found that the further apart that galaxies are from each other, the faster they move away from each other. It was the basis for Hubble's law, which states that the further away a galaxy is from the earth, the faster it is moving away – the distances between galaxies are continuously increasing and therefore the universe is expanding (the theory of universal expansion). Hubble then created the equation of universal expansion that is still used today, although in an updated form. He estimated the expanding rate of the universe at 500 km (311 miles) per second per megaparsec (a distance of approximately 3.26 million light years), meaning that a galaxy 3.26 million light years away from us is receding at 500 km per second. This estimate is known as the Hubble constant: one of the most important numbers in cosmology, used to estimate the size and age of the universe.

It is now thought that Hubble underestimated the distances between the galaxies, which makes his calculation of the expansion rate too large. Astronomers today estimate the number at around 70 km (44 miles) per second per megaparsec, although there still remains a significant amount of uncertainty in the value of the Hubble constant.

The Hubble space telescope, launched in 1990 and named in honour of the great astronomer, aims to provide further data to confirm and refine the Hubble constant. The telescope has so far helped show that the universe is not only expanding, but is expanding at an accelerated pace, driven by a mysterious force dubbed 'dark energy'.

In 1964, cosmic microwave background radiation was discovered and is considered to be an 'echo' of the Big Bang. The Big Bang theory remains the prevailing cosmological view.

Supernovas, Neutron Stars and Dark Matter: Fritz Zwicky

In 1935 Swiss astronomer Fritz Zwicky (1898–1974) used a Schmidt telescope at his mountaintop observatory. This wide-field-of-view telescope was ideal for searching for the ultra-bright stars he named supernovas.

Zwicky hypothesized that a supernova represents the spectacular death of a massive star – a cataclysmic explosion of far greater intensity than a normal exploding star (a 'nova') and visible for only a short period. The star blasts apart, releasing enough energy to outshine its entire galaxy, spewing out particles that will form the foundations of new worlds and leaving behind a crushed remnant, called a neutron star. This stellar remnant is the densest and smallest star known to exist in the universe, composed almost entirely of neutrons or subatomic particles without any overall electrical charge.

The earliest observation of what we now call a supernova was in China in 185 CE. A handful of others were seen before the development of the telescope, and hundreds have since been recorded. Zwicky detected 120, and the hunt for supernovas continues today using computer-controlled telescopes. In each galaxy there are only two or three supernovas per century, but in a universe of a hundred billion galaxies, theoretically there could be thirty supernovas per second!

One of the largest stars in our own galaxy, Betelgeuse, is close to the end of its life and expected to explode as a supernova within the next million years. Watched since ancient times, this bright orange-red 'supergiant', part of the Orion constellation, has used up its supply of hydrogen; the core has compressed and the outer layers have expanded into an overblown star visible to the naked eye.

Cosmic rays, or high-energy radiation, are a side effect of supernovas that can affect electronic devices in satellites. They are possible causes for malfunctions in crashed airliner flight control systems and place a significant barrier to interplanetary travel by crewed spacecraft in the future unless effective shielding can be developed.

In 1933 Zwicky discovered one of the greatest mysteries of modern astrophysics: dark matter. As its name suggests, dark matter is not visible with telescopes but its presence can be inferred from its gravitational effect on stars and other visible matter. His discovery came when he noticed that the mass of stars in the Coma cluster of galaxies would never be enough to pull these galaxies together into a cluster by gravitational force. He concluded that dark matter must exist to make up for the 'missing mass' in the universe. Vera Rubin in the 1970s provided evidence for Zwicky's theory when she noticed a strange discrepancy: the stars at the edges of galaxies move faster than predicted using the law of gravity.

It is now thought that dark matter makes up about 27 per cent of the matter-energy composition of the universe, with dark energy (the unknown force causing the universe's expansion to speed up) making up about 68 per cent and ordinary, visible matter just 5 per cent.

White Dwarves and Black Holes: Chandra

Born in Lahore when it was part of British India (it is now in Pakistan), Chandra, or Subrahmanyan Chandrasekhar (1910–95), was perhaps inspired by his scientist uncle, Sir C. V. Ramen, who won the Nobel Prize in Physics in 1930. Arriving in England for postgraduate work, Chandra moved on to America when his revolutionary ideas aroused opposition and scepticism.

Chandra's most famous theory states that when the nuclear energy source at the heart of a star (such as our sun) runs out and the star approaches its last stage of evolution, it does not necessarily end up as a small, stable, slowly cooling fragment known as a white dwarf. Instead, if its mass is above a certain limit (the 'Chandrasekhar limit', which is higher than the mass that gives birth to a neutron star), it will explode in a supernova, then carry on collapsing into itself to form an infinitely dense, infinitely small point, now known as a black hole. A black hole's gravity becomes so powerful that anything, including light that gets too close, is pulled in.

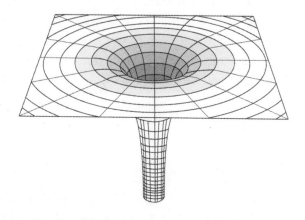

A black hole is a single point of infinite density with gravity so powerful that anything that comes close is pulled inside.

To come to this conclusion Chandra applied rigid mathematics, including the new ideas of quantum mechanics, and the special theory of relativity to the known properties of white dwarf stars.

He surmised that at large masses the Pauli exclusion principle (formulated by Wolfgang Pauli in 1925) – also known as the electron degeneracy principle – would apply. This states that no two electrons can occupy exactly the same quantum space. A consequence of this, according to Chandra, is that the contracting pressure of a massive collapsing star will force electrons to move outwards to higher energy levels at near light speeds. This will bring about an explosion, blowing away the envelope of electron gases surrounding the dying star and leaving the remnant as a dense, still-collapsing fragment.

Chandra's work on the structure, origins and dynamics of stars and his prediction of black holes were later verified.

Pulsars, Quasars and Little Green Men: Susan Jocelyn Bell

Quasars were first discovered in the early surveys of radio waves in the 1950s (the name is short for quasi-stellar radio source). They exist at the edge of the visible universe, some 10–15 billion light years away. When radio waves, light and radiation emitting from quasars reach earth we are effectively looking back 10–15 billion years into the past.

A quasar is believed to be a supermassive black hole surrounded by a flat, spiralling disc-like structure of gas (an accretion disc). It has super-strong gravity and can attract stars and even small galaxies into the black hole, causing a massive output of radiation energy and light in a distinctive flare, which is how quasars are spotted.

British astronomer Susan Jocelyn Bell (born 1943) was monitoring the recently discovered quasars using a basic antenna consisting of wires strung on stakes in a field near Cambridge University when she was puzzled by faint but regular radio pulses. The research team wondered whether 'Little Green Men' were trying to communicate from outer space, but realized that the pulses were coming from spinning neutron stars – the ultra-dense tiny stars composed of neutrons (subatomic particles without an electric charge) that Fritz Zwicky had hypothesized in 1935 as the remains of a star after a supernova explosion (*see* p. 45). They were named pulsars. Bell missed out on receiving a Nobel Prize for her discovery because of her student status at the time, which caused some outrage.

It is believed that up to 30,000 neutron stars inhabit our galaxy. Giant radio telescopes are trained on the skies attempting to pick up their pulsed signals.

Singularities: Stephen Hawking

One of the world's most famous cosmologists, Stephen Hawking, did much to further human understanding of the creation, evolution and present structure of the universe.

Working on general relativity in the 1960s, Hawking and Roger Penrose (born 1931) developed new mathematical techniques designed to show that in the past there must have been a state of infinite density, or Big Bang singularity, where all the galaxies were on top of each other and the density of the universe was infinite.

Prior to Hawking, scientists believed that nothing at all could escape a black hole. Hawking discovered that under particular conditions a black hole can emit certain subatomic particles, known as 'Hawking Radiation'. He also showed that black holes have a temperature and they are not entirely black. They obey the laws of thermodynamics and eventually they evaporate.

Stephen Hawking (1942–2018)

Inspired in part by his father, a specialist in tropical diseases, Hawking was interested in fundamental scientific questions from his early teens. After completing his physics degree at Oxford University he became a doctoral candidate in cosmology at Cambridge, but soon after his arrival there he was diagnosed with amyotrophic lateral sclerosis, the motor neurone disease that produces weakness and muscle wastage. Doctors gave him only a few years to live. Rather than surrender to his apparent fate, this news instilled in Hawking the drive to capitalize on his ability and realize his ambitions to discover the secrets of the universe.

Physical deterioration generated by his condition eventually confined Hawking to a wheelchair. Over the years his speech became slurred and research students would often have to read his lectures on his behalf. In 1985, after an operation that removed his ability to speak altogether, he was fitted with a computer system and speech synthesizer enabling him to deliver public lectures with an electronically generated voice.

Chapter 2

Mathematics: The Science of Numbers

From simple sums to complicated cryptography, the universe is full of mathematical puzzles – how numbers work together, how shapes can be described, how patterns can be made. Unlike most sciences, a mathematical theory can be proven to be a truth and, once proven true, can never be disproven. So the history of mathematics is the development of completely new ideas rather than the replacement of old models.

There are still plenty more puzzles to be solved or, in many cases, questions that have not yet even been properly identified.

The Elements of Geometry: Euclid

Called the 'Father of Geometry', the Greco-Egyptian Euclid (c. 325 to c. 265 BCE) wrote the world's most widely copied secular book, *The Elements of Geometry*. It was used in Europe and the Middle East as the premier maths textbook for about 2,000 years; despite its name, it covered all aspects of mathematics that were then known in a clear, concise and easily understandable format. For some of the mathematical proofs and theorems, no one has been able to provide better explanations than Euclid.

The thirteen-volume *Elements* contains definitions, theorems, proofs and unproven postulates or axioms (statements that are held to be self-evidently true although there is no proof). His work covered both theory and practical applications, making it especially valuable for those who wanted to study and then apply mathematics.

Although we know nothing about the man other than the fact that he taught in Alexandria, a major centre of learning at the time, he gave his name to Euclidean geometry – the study of points, lines, planes and other figures, which he brought together under a common set of assumptions. Among much else, the *Elements* includes famous principles such as the golden ratio (relating to the geometrical basis for beauty), how to construct the five regular solids known as the Platonic solids, and Pythagoras' theorem (covering the squares of the sides of a right-angled triangle). Incidentally, the Greek Pythagoras (sixth century BCE) did not formulate the theorem named after him, but he may have been the first to prove it.

When asked by the pharaoh for a shortcut to understanding maths, Euclid replied, 'There is no royal road to geometry.'

The Mathematics of Machines: Archimedes

Archimedes of Syracuse was known by his ancient Greek contemporaries not so much for his original mathematical thinking, but more for his mechanical devices. Some, such as the water pump known as 'Archimedes' screw', and a compound pulley, had valuable social applications, while others, probably his most notorious inventions, were war machines such as a giant catapult and 'Archimedes' claw', which was also known as the 'ship-shaker'. A crane with a massive grappling hook or claw, this weapon could smash into ships, sink them or even lift them out of the water.

Levers were important features of his machines, and although he did not invent them, Archimedes used the principle of equilibrium to offer the first explanation of how they work. His most famous theorem – Archimedes' principle – describes how to calculate the volume of a body immersed in liquid by measuring the volume of displaced water. He reached this conclusion in order to answer the tricky question of how to prove whether or not the king of Syracuse's new gold crown had been adulterated with cheaper metal.

Archimedes also used the 'method of exhaustion' to find the area of a circle. This involves drawing regular-sided polygons (figures with at least three straight sides and angles) inside and outside the circle, then adding sides to the polygons until they approximate the curve of the circle. (The properties of a polygon are much easier to calculate than those of a circle.) He also discovered relationships between spheres and cylinders, and explored other areas of maths such as square roots and the properties of shapes (triangles, rectangles, circles and so on).

Archimedes (*c.* 287 to 212 BCE)

The son of an astronomer, Archimedes was born in the independent Greek city-state of Syracuse on the island of Sicily. He is famous not just for his mechanical devices but also for jumping out of the public baths and rushing home naked shouting 'Eureka!' ('I have found it!') when he hit upon the theorem known as Archimedes' principle.

He always thought his mathematical theories were more significant than his mechanics. In his book *The Method* (*c.* 250 BCE) he wrote, 'Certain things first became clear to me by a mechanical method, although they had to be proved by geometry afterwards …'

In 218 BCE Syracuse became involved in the Second Punic War (218–201 BCE) as an ally of Carthage against Rome. (This was the same war in which Hannibal took elephants across the Alps to attack Rome.) Archimedes' war machines helped repel the Roman invaders for several years, but Syracuse was overrun in 212 BCE and Archimedes was killed. The grand old man of mathematics was promised safe conduct, but according to one legend, he was so caught up in a maths problem that he ignored the legionaries sent to take him to the Roman general, and in exasperation they ran him through. Another story has it that he was carrying scientific equipment when Roman soldiers killed him for what they hoped was valuable loot.

Pi to Seven Places: Zhang Heng and Zu Chongzhi

Probably the most famous number in the world is pi, the name for the Greek symbol of the number, π.

Approximately twenty-two divided by seven and usually recorded as 3.14, pi describes the properties of a circle; where r is its radius, whatever the size of a circle, its circumference is always $2\pi r$ and its area is always πr^2. So pi is quite useful in applied geometry – in fact, the height of the ancient Egyptian pyramids of Giza was determined by the ratio of pi and the pyramids' perimeters. It is also a crucial element for trying to solve an enduring mathematical problem: using just basic instruments such as a ruler and compass, can a square be constructed that is equal in area to a specific circle?

No one person discovered pi first. It was calculated independently in every early civilization that had a mathematical science: Babylon, Egypt (as mentioned), Greece, India, China, the Maya of Central America and others. Using several different geometrical approaches, most early mathematicians reached a value of between 3.12 and 3.16. One Chinese inventor, Zhang Heng (78–139 CE) proposed that it was the square root of ten: 3.162.

But it was Zhang's later compatriot, the astrologer, engineer and mathematician Zu Chongzhi (429–500 CE), who was the first person in the world to calculate pi correctly to seven decimal places, between 3.1415926 and 3.1415927. It would take a thousand years before such accuracy was replicated in Europe.

Zu's main interest was calendar reform, and he was the first Chinese calendar-maker to take into account the precession of the equinoxes (*see* p. 18). His calendar was amazingly accurate, calculating a year as 365.24281481 days, just fifty seconds out from today's calculations.

Zu did not live to see his calendar adopted in China but during his lifetime he became famous for his inventions, such as a south-pointing chariot and a paddle-wheel boat. Another of his legacies was a book on mathematics that had to be dropped from the imperial syllabus since it was too difficult for most scholars!

Pi itself continued to be a fruitful source of mathematical ponderings. In 1882, Ferdinand von Lindemann (1852–1939) showed that it is a transcendental number: infinite and without a predictable pattern. And in 2011 a computer programme took 191 days to calculate 10 trillion decimal places of pi. No doubt someday a computer will give us a google's worth of decimal places, but it will get us nowhere nearer constructing our square.

Tables of Sines: Aryabhata

Ancient India had a sophisticated mathematical culture, which was revived by the young genius Aryabhata (476–550 CE). His important book, the *Aryabhatiya*, was written when he was just twenty-three.

In that book, written in 119 poetic verses, Aryabhata became the first person to give a method of finding square roots and to outline the basics of trigonometry, later called tables of sines. One of his methods to create his tables used the Pythagorean theorem. He also showed how to project points and lines on the surface of a sphere onto a plane, thereby applying plane trigonometry to the geometry of a spherical shape.

Aryabhata offered innovations for algebra and astronomy, but two of his most significant concepts were his use of decimal place values indicating tenths, hundredths, thousandths and so on, and his understanding of the mathematical concept of zero. Although all early civilizations would have realized that having no harvest results in hunger, zero as a mathematical idea allows the development of negative numbers and is an important stage in elementary arithmetic and the growth of mathematics as an intellectual pursuit. Through his work these concepts were transmitted to the Middle East, built upon and developed, and then taken to Europe.

The Decimal Point Enters Europe: Fibonacci

It seems extraordinary that Western Europe had no mathematical concept of zero until 1202. That was the year that the young Italian accountant Fibonacci (1170–1250) published his groundbreaking work *Liber abaci*, which introduced several crucial Indian-Arab ideas to Europe. These included Arab numerals, the mathematical idea of zero and the decimal numbering system with place values.

Fibonacci was actually called Leonardo of Pisa, but he is best known by the name that means 'Son of Bonacci'. It was

Bonacci, a mercantile agent, who suggested his son should study Arab mathematical concepts as part of his business training in North Africa. Returning to Italy, Fibonacci persuaded Europeans that the Arab system was much simpler than Roman numerals and offered more accurate calculations. Use of the mathematical zero led to the concept of negative numbers, or those less than zero, so Fibonacci also laid the groundwork for the future development of number theory in Europe.

A sophisticated mathematician who found practical applications for abstract theorems, Fibonacci wrote books that were particularly useful to merchants since many of his examples were to do with business, such as how to calculate costs and profits or how to convert between the major currencies of the Mediterranean. He also offered solutions to surveying problems.

Fibonacci is best remembered for asking how many rabbits can be produced under certain circumstances. His answer, summing each two preceding numbers, is known as the Fibonacci sequence and can be applied in many areas of science, mathematics and nature. Although in *Liber abaci* he actually omitted the first term, the sequence starts as 1, 1, 2, 3, 5, 8, 13, 21, 34, 55, and so on, with (apart from the very beginning) each number being the sum of the previous two. When squares are created from these numbers, a spiral can be formed by connecting the opposite points of each square. Number patterns such as this are beloved by mathematicians, but the Fibonacci sequence has also had practical use in helping solve some maths problems. It interests other scientists because it is used in some computer software, it describes part of a model of economic growth and it is found in several natural objects, for example the way some leaves branch on a stem, the whorls of pineapples and pine cones, the arrangement of sunflower petals and the distribution of seeds in a raspberry.

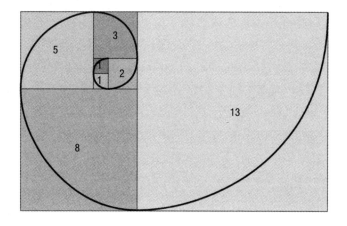

A visual representation of the Fibonacci sequence.

Cartesian Coordinates: René Descartes

With his invention of Cartesian coordinates, the French philosopher–mathematician René Descartes was responsible for generations of schoolchildren puzzling over their x and y axes on graphs.

Descartes was idly watching a fly wandering around the walls and ceiling when he realized that the fly's progression could be represented both geometrically – by the line of its path and the shapes that were created by the line – and algebraically, by a series of points. He then drew up a Cartesian plane (named after the Latinized version of his name, Cartesius), using numbered, perpendicular vertical and horizontal lines, or axes, on a flat surface to describe the position of points.

He developed this idea at about the same time as his fellow Frenchman Pierre de Fermat (*see* p. 64), leading to a lively debate over priority. As a result, the two greatest mathematical minds of their time never collaborated.

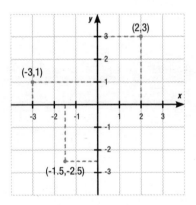

A four-square grid of Cartesian coordinates. The two number lines are at right angles to each other and meet at a point called the origin, with the lines above and to the right of the origin measured in positive numbers, while the lines below and to the left are measured in negative numbers. Every point on the grid is referenced by two coordinates, the distances from the origin along each axis, with the horizontal measurement listed first. The coordinates are noted in brackets, with the origin being (0,0).

Apart from simple graphs, map references are a common, everyday use of Descartes' invention, but in the mathematical world this new contribution to 'analytical geometry' provided a groundbreaking link between algebra and geometry. It allowed algebraic terms to be expressed as coordinates or lines and, vice versa, geometric shapes to be expressed as algebraic equations. It was a foundation for the later discovery of infinitesimal calculus by Isaac Newton (*see* p. 69), who was deeply influenced by Descartes' ideas.

Descartes also championed the notation, which is standard today, of using superscripts to indicate exponents or powers, for example 2^{10}.

René Descartes (1596–1650)

Descartes formulated his theory of analytical geometry while sweltering in an overheated room. During this episode he experienced visions that guided him to try to connect logic and philosophy.

As the 'Father of Modern Philosophy', he is well known for his statement '*Cogito, ergo sum*' ('I think, therefore I am'), which he reached by his method of doubting everything else. He was careful 'never to accept anything for true which I did not clearly know to be such ...'

Descartes wrote one of his most important works, *Discourse on the Method*, in French rather than the scholarly language of Latin so that everyone (even women, he pointed out) could read his work.

He is one of the few mathematicians to have a place named after him: the town in which he was born, La Haye en Touraine, was renamed Descartes in his honour.

Number Theory: Pierre de Fermat

Since the days of the ancient Greeks, the branch of mathematics known as 'number theory' was neglected in the West until French lawyer Pierre de Fermat (1601–65) revived the discipline. Sometimes called higher arithmetic, it is concerned with the properties and relationships of numbers, and Fermat was the first to work solely with whole numbers. He would not accept fractions in a solution to any of the problems that he set.

At about the same time as René Descartes (*see* p. 61), Fermat developed a coordinate system, so he is one of the founders of analytical geometry. He also collaborated with Blaise Pascal, founding probability theory (*see* p. 65).

Fermat followed François Viète (1540–1603) in believing that algebra could be used to analyse mathematical problems. He was also inspired by ancient Greeks such as Diophantus, whose book *Arithmetica* posed conjectures and stated theorems, leaving it up to the reader to find proofs or solutions. Similarly, Fermat set mathematical puzzles, and he seldom explained how he himself had reached a proof.

In the long term, Fermat directly influenced modern number theory. But in the short term his impact was lessened because he seldom published his work, which is mainly known through his correspondence with other learned men of the day or as marginal notes in books. He is best known for his little and last theorems. The first, a tool in the ongoing search for prime numbers, states that in $n^p - n$ where p is a prime, the result will always be a multiple of p. The latter was 'last' because all his other puzzles were solved long ago, yet a solution was not found for this one for nearly 300 years (*see* p. 76).

Projective Geometry and Probability Theory: Blaise Pascal

In the seventeenth century the French genius Blaise Pascal helped take mathematics in two new directions: projective geometry and probability theory. He also built one of the first mechanical calculating machines and gave his name to an intriguing number pattern, Pascal's triangle.

Pascal was only sixteen when he presented Pascal's theorem of projective geometry. This studies the relationships between geometric figures and the images that occur when they are projected onto another surface. His theorem showed that if a hexagon is drawn inside a conic section then the three opposite sides of the hexagon will intersect at points on a straight line (the Pascal line, *see* p. 66).

In 1654 Pascal corresponded with Pierre de Fermat (*see* p. 64) over two gambling problems: how often to expect a double six, and how to divide the stakes fairly if the gamblers finish a game early. The result was probability theory, introducing the expected value of a variable in specific circumstances.

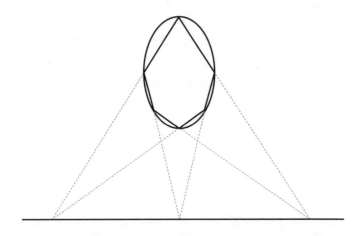

A hexagram in a conic section showing Pascal's theorem.

Pascal's famous calculating machine – called a Pascaline – was developed to help his father, who was a tax collector. The machine added and subtracted, but although it was a forerunner of modern computers it was a commercial failure because it was expensive and awkward to use.

As well as his considerable mathematical achievements, Pascal not only developed the law of pressure known by his name and invented the hydraulic press and the syringe, but also proved the existence of vacuums. Fellow French thinker René Descartes (*see* p. 63) simply could not accept the possibility of vacuums and wrote in response that Pascal 'has too much vacuum in his head'.

Blaise Pascal (1623–62)

Blaise Pascal was educated at home by his father, who believed that the boy should not approach mathematics until he was at least fifteen. So to begin with he was entirely self-taught in that area.

In 1646 Pascal adopted Jansenism, a movement within Roman Catholicism that was considered by some to be heretical. He became increasingly religious, and in 1654 interpreted a vision as meaning that he should turn his back on the world and instead adopt a life of prayer. From then on he did very little mathematical work, and spent the years before his death from cancer in 1662 writing religious reflections, his *Pensées*. These included 'Pascal's Wager', which gives a probabilistic argument that it is rational to believe in God since: 'If God does not exist, one will lose nothing by believing in him, while if he does exist, one will lose everything by not believing.'

Binary Numbers: Gottfried Wilhelm von Leibniz

The German polymath Gottfried Wilhelm von Leibniz (1646–1716) is famous for becoming embroiled in the 'Calculus Precedence Controversy' with the Briton Isaac Newton (*see* p. 70), and for many years British scientists refused to acknowledge Leibniz's contributions to science and mathematics.

Apart from calculus, Leibniz influenced another major part of the modern world by refining the binary system, which ultimately paved the way for the digital revolution and our computers. In 1679 he published the clearly titled *Explanation of Binary Arithmetic*, in which he laid out the system as it is used today.

Binary is essentially a counting system in base two, that is, using only two digits instead of ten, but before Leibniz letters were used for the two characters. He introduced the digits 0 and 1, and set out the representation of the system reading from right to left.

Counting in binary quickly becomes cumbersome to most people, but it is the standard basis of digital equipment, which is now as essential to us as our two hands.

A lawyer, courtier and diplomat, Leibniz was also a philosopher, holding optimistic views that the world is the best that God could create. He still found time to write on topics as varied as a library cataloguing system and symbolic logic. Leibniz was parodied by the French author Voltaire, who championed Newton, as Professor Pangloss in the 1759 novella *Candide*.

Calculus: Sir Isaac Newton

The English 'natural philosopher' Isaac Newton is best known for his theory of gravity and his laws of motion, but his work in physics and mathematics cannot easily be separated. His influential 1687 tome *Principia* explored many aspects of maths as well as introducing his view of the universe. He even provided a mathematical equation for the force of gravity:

$$F = \frac{G\, m_1 m_2}{r^2}$$

In 1665, Newton first began to work on what would be a revolutionary mathematical development – calculus, or the study of change. In particular, he wanted to calculate the changing speeds of falling bodies and planetary orbits at specific moments in time.

His approach used what he called 'fluxions', which are algebraic expressions of tangents, the precise slope of a curve (such as orbits) at any particular point. These let him calculate the magnitude of 'flow' or change along the curve, and unveiled that a derivative function gives the slope at any point of a function. He also discovered that the rate of change is inversely related to the integration or summation of the area bounded by a curve.

Calculus is an essential tool for advanced mathematical analysis since it enables calculation of both areas (integral calculus) and changes in a system (differential calculus). One of its applications is to set minimum credit-card payments at the exact time of issuing the statement.

Newton engaged the German Gottfried Wilhelm von Leibniz (*see* p. 68) in a lasting argument about precedence and plagiarism over calculus. But the two probably invented calculus independently, if simultaneously, and they certainly used different methods. While Newton focused on deriving a function, Leibniz integrated a function to calculate areas and volumes. It is his notations that are now used in mathematical calculus: ∫ for summing areas or integration, and *dy/dx* for differentiation or rates of change.

Calculus was not Newton's only contribution. He made advances in so many mathematical problems (the generalized binomial theorem, Newton's method of approximating the roots of a function, power series and classification of cubic curves) that he practically laid out a map indicating the paths that future mathematicians would take.

But in 1676 he wrote, 'If I have seen further it is by standing on the shoulders of giants.'

Sir Isaac Newton (1642–1727)

One of the greatest scientific figures of all time, Isaac Newton (*see* p. 40 and p. 69), from Lincolnshire, England, formulated many of his ideas in 1665–6 when he was staying at his family home because his university,

Cambridge, was closed because of plague. It was then that, according to legend, he was hit on the head by a falling apple and conceived his theory of gravity.

Those two years saw him make extraordinary scientific advances, although he did not publish his ideas for many years. He was hugely sensitive to criticism, and when in 1671 his early theories on light and colour were not well received, he withdrew into a private study of alchemy. He had to be cajoled into publishing his great work, *Principia*.

Newton also wrote extensively on alchemy, ancient history and Bible studies. He became a Member of Parliament, reformed the Royal Mint, and from 1703 onwards was elected annually as President of the Royal Society. He was knighted in 1705.

Fundamental Theorem of Algebra: Carl Friedrich Gauss

Gaussian curvatures (of surfaces), Gaussian probability distribution, the gauss as a unit of magnetic field strength – the contributions of the German polymath Carl Friedrich Gauss to mathematics and science (*see* p. 72) were so numerous that they earned him the title 'Prince of Mathematics'. Before he was twenty, Gauss achieved one of the greatest advances in geometry since the ancient Greeks when he proved that a regular seventeen-sided polygon could be constructed using a ruler and compasses.

In 1799 he offered another major achievement – a proof for the 'fundamental theorem of algebra'. Despite its name, the theorem is not actually fundamental to modern algebra, and was one of many puzzles proposed by early mathematicians. Gauss created algebraic curves from expressions of an equation and analysed the curves using topology, a form of geometry studying properties that do not change even if angles and lines do. He created his proof by extrapolating the relationships between his curves with a circle.

In 1801 he published *Disquisitiones Arithmeticae* (Arithmetical Disquisitions), which was the first systematic textbook on algebraic number theory or 'higher arithmetic'. He summarized the scattered writings on the subject, presented his own theories on outstanding problems, and laid down the definitive analysis of concepts and research areas. Among many other advances, he introduced the symbol ≅ for congruence.

He once described number theory as having a 'magical charm' that meant 'it so greatly surpasses other parts of mathematics'.

Carl Friedrich Gauss (1777–1855)

Born in Brunswick, now in Germany, Gauss was a child prodigy from a working-class family – his mother was illiterate and did not record the day of his birth. When he was fourteen his mother and teachers brought him to the attention of the Duke of Brunswick, who provided the genius with a stipend that allowed him to remain at school and go to university at Göttingen.

By 1801 he had established his basic mathematical and scientific approach: intense empirical investigations, followed by reflection, then the construction of a theory. He studied and mastered an astonishing range of subjects, from astronomy to surveying and magnetism, let alone several areas of mathematics. His habit was to investigate a field of study, make a discovery or invention, then move on to another subject. He was a true polymath.

But in his personal life he did not like change, so he did not embark upon the lecture tours that would have found him public recognition. He produced 178 publications, and in addition left a host of unpublished papers, notes and memoirs.

The Three-Body Problem and Chaos Theory: Henri Poincaré

Isaac Newton (*see* p. 70) was just one of the great minds who tried and failed to give a mathematical equation to show how the solar system worked. Specifically, the maths explaining the motion of more than two orbiting bodies – and why they never collide – proved elusive until 1887 when King Oscar II of Sweden set up a prize for the solution to this three-body problem, now known as the *n*-body problem (where *n* is a number greater than two).

The French mining engineer and mathematician (Jules) Henri Poincaré (1854–1912) was already exploring the complex differential equations that describe the solar system's stability. He reduced the problem to a simpler form, looking at two large bodies and a third that was so much smaller that it had no gravitational effect on the others. He was thus able to show that the smaller body has a stable orbit, but he could not prove that that orbit might not swing it far away from the other bodies.

His contribution was significant enough that he won the prize, but he then uncovered an error meaning that the orbit could be totally chaotic: the smallest change could result in larger, unpredictable motions. He had accidentally discovered chaos theory.

For lack of computing power, as we would say now, the study of chaos theory became dormant until the 1960s, when computers made it possible to calculate the many permutations that result from making tiny changes to a system. Then, American meteorologist Edward Lorenz (1917–2008) applied it to his models of weather changes, coining the term 'butterfly effect'.

The *n*-body problem has still not been completely solved.

Artificial Intelligence: Alan Turing

Fascinated by mathematical logic, English codebreaker Alan Turing devised a test for whether electronic digital computers could simulate human intelligence. He called this the 'imitation game'.

In 1950 he published his 'Turing Test' in a paper entitled 'Computer Machinery and Intelligence'. The test required three participants – a human, a machine and an interrogator – previously unknown to each other, sitting in different rooms but in contact via teleprinter communication. Modelling the human mind as a physical machine, he produced a series of tests designed to see whether the responses of the computer were indistinguishable from the human's responses. Turing's paper also presented a series of arguments against the claim that machines cannot display human intelligence.

In 2014 a Russian computer program passed the test by convincing more than 30 per cent of its human interrogators that it was a thirteen-year-old boy.

Turing's research on machine intelligence raised many important philosophical questions about artificial intelligence and human consciousness. In 1950 he wrote, 'I believe that at the end of the century... one will be able to speak of machines thinking.' It is still science fiction, but maybe not for long.

Alan Turing (1912–54)

Born in London, Alan Turing was one of the key figures in the development of computers with his 1936 description of a hypothetical machine for automatically performing functions that were fed into it.

In 1938 Turing attended the British government's Code and Cypher School, so when the Second World War broke out the following year he was perfectly placed to join a team based at Bletchley Park that was working to crack Nazi codes. These codes were encrypted by the German machine known as Enigma, and Turing's wartime effort resulted in the cryptanalytic machine called Bombe, which eventually cracked the cypher. It was a major contribution towards the Allied victory, and the theory of information and statistics that he advanced helped make cryptanalysis into a science.

In 1952 Turing was arrested for a homosexual act, which was illegal in those days. His high-level security clearance was revoked, and in 1954 he committed suicide by eating a cyanide-poisoned apple.

Solving Fermat's Last Theorem: Andrew Wiles

English mathematician Andrew Wiles (born 1953) first wondered about Fermat's last theorem (*see* p. 65) when he was just ten and came across the 326-year-old problem in a library book. It was to be thirty more years before he found a proof for the theorem.

Pierre de Fermat (*see* p. 64) presented this puzzle in 1637 in a scribble in his copy of the ancient Greek Diophantus' book *Arithmetica*, where he noted that his proof (a 'truly marvellous demonstration') would not fit in the margins. Wiles used techniques that were not available to Fermat, so many mathematicians now think the Frenchman was mistaken in claiming he had a proof.

Fermat's last theorem states that the simple equation $a^n + b^n = c^n$ can only be solved with positive integers (whole numbers to you and me) if n is no greater than two.

Fermat did note that $n = 4$ is a special case that is easily solved, so the challenge does not apply in this instance. By the mid-nineteenth century the theorem was proven for many prime numbers, and with computers it became possible to do the calculations that proved the theorem for all primes up to 4 million. But a proof for all numbers was considered to be 'inaccessible' – impossible or at least not possible with current knowledge.

Nevertheless, successive work by twentieth-century mathematicians showed in 1986 that the theorem could be connected to the Taniyama-Shimura-Weil conjecture (later known as the modularity theorem), which linked elliptic curves to modular forms – complex analytic functions in *four* dimensions. If the link was correct, any solution to the Fermat equation would create a non-modular elliptic curve, so could not exist. This, along with other new ideas, revived Andrew Wiles' interest in the problem.

In 1994 Wiles had succeeded in proving enough of this new modularity proposal that it also proved Fermat's last theorem. However, his first proof was found to have a minor error. Working with former student Richard Taylor (born 1962), Wiles circumvented this problem, and published his completed proof in 1995.

In 2000 Wiles received a knighthood for his achievement in solving this enduring mathematical puzzle.

The World Wide Web: Tim Berners-Lee

Computers developed from early counting machines such as the Pascaline (*see* p. 66), the difference engine of Charles Babbage (1791–1871), the algorithms of Ada Lovelace (1815–52) and the Turing machine (*see* p. 76). Linking computers together in networks showed how powerful shared computers could be, and Tim Berners-Lee (born 1955), an English computer scientist, took this a stage further in 1989 by inventing the World Wide Web.

Berners-Lee wrote his initial proposal for the Web while working at CERN, the European Organization for Nuclear Research. He envisaged a global information space where computers were linked in a vast network and data was freely available to all. At this time the internet existed as basic computer-to-computer networks used by scientists and the military, but Berners-Lee realized that by using hypertext links that allowed a computer user to 'jump' to another document, he could create a web of documents accessed through the internet.

In 1990 Berners-Lee brought his vision to fruition by writing the Hypertext Transfer Protocol (HTTP), the language used by computers to transmit hypertext documents. He also wrote HTML, the Hypertext Markup Language with which to format hypertext pages, developed a client program or browser to enable the pages to be viewed, and set up the first Web server.

He declined to patent his invention, since he always wanted it to be available for everyone, and has campaigned to keep all areas of the Web open to all. In 1994 he set up the World Wide Web Consortium (W3C), which oversees the Web's standards and development. In 2004 he was knighted, and he has also received honours from universities and institutions around the world.

Young people today can barely imagine a world where their computers – let alone their phones or tablets or even smartwatches – are not linked to others around the world. Berners-Lee's invention has truly revolutionized communications and the flow of information.

CHAPTER 3

Physics:
What Things are Made Of

THE WORD PHYSICS derives from the Greek for 'nature', and this science explores the nature of all things. The laws of physics are the laws of nature, and a major part of the history of this science is the search for common laws that apply to all parts of the universe equally, from a star to an atom. Today, physics particularly refers to the study of matter and energy, or particles and the forces that act on them.

Until the closing decades of the nineteenth century the physical world was entirely explained according to the principles of classical (or Newtonian) mechanics: the physics of everyday life. By about 1900, however, there were whole new areas of study – relativity and quantum physics – where Newtonian mechanics no longer applied.

So physics is now divided into two sections. While it is normal in science for modern ideas to replace older ones, 'modern' physics did not supersede classical physics, but instead sits side by side with the older concepts. Classical physics applies to the world that we experience, for example sound, electricity and machines. Modern physics, such as quantum mechanics, particle physics or relativity, deals with the extremes of nature: the smallest known particles of an atom, light speeds or very massive objects.

Early Theories of Elements and Particles: Thales and Aristotle

Physics is often said to have begun with the ancient Greek philosopher Thales of Miletus, who was born around 624 BCE. He was the first known person to argue that superstition and faith should be put aside, and that one should explain natural phenomena in terms of observed facts. Unfortunately he observed a lot of water, and theorized that the whole world is made from water in different forms.

Thales was forgotten for many centuries, and instead Western scientists followed the theories of the great Greek scientist/philosopher Aristotle (*see* p. 17). He believed that everything on earth was made up of four elements – earth, air, fire and water.

Aristotle's ideas were adopted into Christian philosophy, so in the early part of the European Middle Ages it was not considered appropriate to question his view of the universe. It was not really until the Renaissance that European science flourished. One of the first great Renaissance minds was Leonardo da Vinci (1452–1519), artist, mechanical engineer, inventor and all-round scientist.

Newtonian Mechanics: Sir Isaac Newton

It is a myth that an apple fell on Isaac Newton's head (*see* p. 71), causing him to 'discover' gravity. But it is true that he formulated his theory while wandering in the garden and musing on why apples always fall directly to the ground. He published his groundbreaking theory of gravity in 1687 in one of the world's most important scientific books, *Principia*. This also contained his laws of motion, which formed the basis for classical mechanics. If that was not enough, Newton (1642–1727) was also the co-inventor of mathematical calculus (*see* p. 69), made important breakthroughs in optics, and helped develop the modern scientific method of investigation, experimentation and analysis.

Newton's three laws of motion explain how forces and masses interact to create movement. They are:

I. Every object in a state of uniform motion tends to remain in that state of motion unless an external force is applied to it. This is also called the law of inertia, and confirmed the ideas of Italian scientist Galileo Galilei (*see* p. 38), who had studied pendulums and falling bodies.
II. The relationship between an object's mass m, its acceleration a and the applied force F is $F = ma$. This simple equation allows calculations of dynamics. It shows why velocities change when affected by forces, and explains why the same force will barely budge a massive object, but can accelerate a less massive object much more quickly.
III. For every action there is an equal and opposite reaction. Here Newton explained so many movements, from

swimming (you push the water backwards away from you and the water reacts, propelling you forwards) to cars skidding on ice (the wheels are unable to exert a force on the ground, which is therefore unable to give a 'reaction' and move the car forwards).

Together, his laws founded classical mechanics, and Newton's theories still underlie classical physics today.

Lightning Under Control: Benjamin Franklin

Many of our words about electricity were coined by Benjamin Franklin: battery, charge, conductor, even electrician. Electricity had long been known as a static phenomenon, and in the 1740s electric machines that could cause friction sparks off amber or other materials were used for entertainment or as parlour tricks.

Franklin believed that the electrical sparks he saw in his laboratory were connected to lightning, and he also believed that electricity did not have to be just static, but was like a fluid in some ways so it could be directed to follow a chosen path. This may have prompted his supposed hair-raising experiment of flying a kite attached to a key in a thunderstorm.

Franklin did indeed establish that lightning and electricity are identical, and invented a lightning rod with a metal cable running up the side of a building, the bottom end buried in the earth and the top attached to a metal rod sticking up into the air.

He was a prolific inventor and is also credited with the design of a heat-efficient stove and bifocal glasses, which he needed for himself.

Benjamin Franklin (1706–90)

Born in Boston, Massachusetts, which was in those days a British colony, Franklin was a leading figure in the American War of Independence, helping to draft the Declaration of Independence. It was he who signed the peace treaty with Britain at the end of the war. His oldest son, William, remained loyal to England, causing a permanent rift between father and son.

Franklin had many successful careers: at different times he was a printer, journalist, postmaster general, diplomat and politician, quite apart from being a scientist and inventor. He was also a philanthropist, helping to found several institutions that still exist, such as the Pennsylvania Hospital and the Philadelphia Union Fire Company. He was a passionate advocate for the abolition of slavery, and he considered that his work in public service was more consequential than his scientific contributions.

Frogs' Legs and the Voltaic Pile: Alessandro Volta

In 1786 the Italian physicist Luigi Galvani (1737–98) observed that recently cut frogs' legs hung up on copper hooks jerked and contracted when they touched an iron railing. It looked like an electric circuit, even though there was no electrical machine in operation. Galvani believed the legs were discharging electricity and that he had seen some form of 'animal electricity'. He theorized that the force was stored within animals' bodies.

His fellow Italian Alessandro Volta (1745–1827) disagreed. Volta's own experiments with frogs' legs and even his own tongue showed that nerves and muscles would convulse when placed between a metallic circuit, and led him to think that any damp material placed between two metal plates would create a continuous current. This paved the way for the battery.

Volta's first 'voltaic pile', as it was called, was a cylindrical stack of zinc and copper discs separated by paper or leather and damp cloth, soaked in salt solution or dilute acid. Wires protruded from the top and bottom. By testing his pile on several substances, Volta proved that it did indeed produce an electric current.

He did little more with his pile, but it was quickly adopted by other scientists as a means of reliably producing electricity. It was soon demonstrated that a chemical reaction in the pile produced the current, and the machine was adopted by Humphry Davy (*see* p. 117) and others for separating chemical substances, or electrolysis.

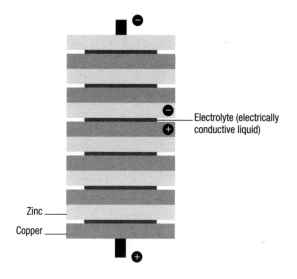

The voltaic pile, the first electric battery.

Atoms, Molecules and Electrons: Amedeo Avogadro and J. J. Thomson

In 1803 the English chemist John Dalton (*see* p. 118) proposed that elements are made up of tiny, irreducible units he called atoms. This was harking back to a theory once proposed in ancient Greece but then dropped in favour of Aristotle's world view. Looking ahead, the atom was going to become crucial for the new particle physics; chemistry and physics were beginning to borrow more and more ideas from each other.

In 1811 the Italian mathematical physicist Amedeo Avogadro (1776–1856) introduced the world to his new 'molecule' concept for groupings of more than one atom.

Avogadro's law about the number of molecules in a volume of gas went unacknowledged for a long time, but his new word was adopted into science.

It took to the end of the century to find the first subatomic particle, which was discovered in 1899 by Englishman J. J. Thomson (1856–1940). He was rerunning an electrical experiment on cathode rays, which were mysterious rays that acted in a vacuum, like electromagnetic waves, yet had some properties of metals and gases. They were explored through cathode tubes or diffusion tubes, and Thomson found that the rays were being attracted to the positively charged side of the electric field, which meant they must have a negative charge, since opposites attract. By now physicists knew that light has no charge so the rays had to be tiny particles. Using new magnetic techniques to weigh the particles, he found them to be 1,800 times lighter than hydrogen, the lightest atom, so they had to be subatomic particles, which he named electrons. He proposed that the electric field was actually tearing them off their atoms inside the cathode tube.

Magnetism: Carl Friedrich Gauss

As well as being an outstanding mathematician (*see* p. 71), the German polymath Carl Friedrich Gauss invented the heliotrope while overseeing a land survey, and invented a magnetometer while helping Alexander von Humboldt (1769–1859) to map the earth's magnetic field in 1832.

Only twelve years earlier, in 1820, the link between electricity and magnetism had been made in Denmark by

Hans Christian Ørsted (1777–1851), who noticed that a compass needle left lying around became magnetized when he turned on his voltaic pile. He thus accidentally confirmed the findings of Frenchman André-Marie Ampère (1775–1836), who showed that the polarity of the magnet can differ depending on the direction of the electric current. Today most of our electricity supply is generated by electromagnetism in one form or another.

Gauss' gadget consisted of a bar magnet suspended from a gold fibre, which he used to measure the strength and direction of the magnetic field in a particular place. With his colleague Wilhelm Weber (1804–91), he built the first electromagnetic telegraph, transmitting messages over 1.5 km (0.9 miles). Just as importantly for him, he arrived at several mathematical principles from studying magnetism and wrote three papers on the subject.

Almost as a side-thought, he also supplied an empirical definition of the earth's magnetic force, produced an absolute measure for it, showed why there can only be two poles and proved a theorem relating magnetic intensity to inclination.

Magnetic Induction of Electricity: Michael Faraday

Formerly an apprentice bookbinder, the English researcher Michael Faraday (1791–1867) first created an electric motor in 1821, to the fury of his mentor Humphry Davy, who resented the apprentice stepping into the spotlight. It was only after Davy's death that Faraday returned to his work on electromagnetism. First he proved electromagnetic rotation. He showed that an electrically charged hanging wire will rotate around a securely fixed magnetic bar, and that a magnetic bar fixed at only one end will rotate around a fixed, electrically charged wire.

His second experiment was to see what would happen if he passed electricity through a coil of wire connected to another one by an iron ring. The second coil also had a wire suspended over a compass. As expected, the first coil of wire became magnetized, but Faraday noticed a flicker of movement in the compass needle as well. He had discovered magnetic induction – the induction of electricity by means of magnetism.

Through other experiments Faraday generated magneto-electric induction, the magnetism being converted into electricity, and proved that there is only one type of electricity. He also discovered the first law of electrolysis: that the chemical effect produced by an electric current on a substance is always proportional to the amount of electricity that flows. He then built an instrument to measure electricity (a voltameter) and used it to prove the second law of electrolysis: that the electrochemical equivalent (electrical charge) of a substance is proportional to its ordinary chemical equivalent.

Faraday's pioneering work introduced terms still used today, such as electrode, anode and cathode. He became a public figure and was consulted on all manner of science issues including the preservation of paintings at the National Gallery in London.

Electromagnetic Radiation: James Clerk Maxwell

A Scottish genius, James Clerk Maxwell (1831–79) is held to be one of the greatest scientists ever. His interests ranged from light (his idea that light is a form of electromagnetic radiation helped Einstein reach the theory of relativity, *see* p. 41) to the application of statistics in physics and physical chemistry. He also produced the first colour photograph.

Before his work on electromagnetism, electricity and magnetism were understood in terms of particles exerting forces upon one another. He showed that they should be understood instead in terms of space-filling fields defined by the Maxwell equations.

These are:

1) Unlike charges attract each other; like charges repel (also called Coulomb's law).
2) There are no single isolated magnetic poles (if there is a north pole there will always be an equivalent south).
3) Electric currents can cause magnetic fields.
4) Changing magnetic fields can cause electric currents.

Maxwell showed that electric and magnetic effects are distinct manifestations of a single electromagnetic force, thus unifying them forever under the electromagnetic field. He described light as 'an electromagnetic disturbance in the form of waves propagated through the electromagnetic field according to electromagnetic laws'.

This was a completely new way of looking at forces, but Maxwell predicted that other 'disturbances in the electromagnetic field' or forms of electromagnetic radiation would exist. Sure enough, other waves such as radio waves and X-rays were discovered. Scientists, whether physicists, chemists or biologists, soon grew used to discussing phenomena in terms of electromagnetic wavelengths.

James Clerk Maxwell was one of the few scientists who fully understood the thermodynamic chemistry approach of Josiah Willard Gibbs (*see* p. 126). He died from cancer at the age of forty-eight.

Radio Waves: Heinrich Hertz

Had German physicist Heinrich Hertz (1857–94) lived for a few more years, he would have seen the world practically transformed by his discovery of radio waves, but he died in his thirties of bone disease.

In the mid-1880s Hertz began experiments to try to detect electromagnetic waves. He used a simple tabletop apparatus, which included a circuit containing an induction coil, a wire loop and a spark gap. At the other end of the table, he arranged another circuit with only a spark gap. He then observed that a discharge from the induction coil across the first gap was accompanied by a weaker spark across the gap in the receiving circuit, thus proving the existence of electrical waves.

Performing further experiments on these waves, later known as radio waves, Hertz showed that they could be reflected, refracted and diffracted.

Hertz's equipment only allowed detection of radio waves across a distance of about 18 metres (60 feet), but he laid the groundwork for the radio communication later developed by Guglielmo Marconi (1874–1937), who transmitted a radio message across the Atlantic Ocean in 1901.

X-rays: Wilhelm Röntgen

In 1895 the German Wilhelm Röntgen (1845–1923) was investigating the properties of cathode rays (electron beams) that were emitted by high-vacuum discharge tubes. While experimenting he discovered that a photosensitive screen lying on his workbench became fluorescent and emitted light when the tube was operating. Objects placed between the tube and the screen cast shadows that could be recorded on photographic plates. The denser the object, the darker the image, so when a hand was positioned there, the bones cast darker shadows than the flesh.

Moving the screen to an adjacent room, Röntgen found that it still produced a luminescent flow when the tube was activated. This high power led Röntgen to conclude that the radiation was entirely different from cathode rays. Since he was unable to establish the exact nature of the newly discovered rays that came from the glass wall of the tube, he called them 'X-rays'. Before announcing his discovery, he conducted further experiments, establishing that X-rays pass unchanged through cardboard and thin plates of metal, travel in a straight line, and are not deflected by electric or magnetic fields.

When Röntgen publicly announced his discovery of these amazing new rays he illustrated his lecture with an X-ray photograph of a man's hand.

X-rays were being used in hospitals within weeks of Röntgen's announcement. They have transformed medical science, and they are also widely used in crystallography, metallography and atomic physics. Not everyone loved them, however. A publication in the USA ran a poem ending:

> For nowadays,
> I hear they'll gaze,
> Thro' cloak and gown – and even stays,
> Those naughty, naughty, Roentgen Rays.

Röntgen refused to benefit financially from his discovery, believing that it should be freely available to all.

Quantum Theory: Max Planck

At the end of the nineteenth century physicists could not explain why the spectrum of radiation emitted by a so-called blackbody did not match expectations according to standard electromagnetic theory. Blackbodies are objects or surfaces that absorb all energy radiated upon them. A surface covered with the pigment lampblack is an almost perfect blackbody, and stars and planets are often modelled as if they are blackbodies.

After some years studying this problem, the German theoretical physicist Max Planck (1858–1947) identified the first demonstrable failure of classic physics models when he proposed that the unexpected spectrum could only be explained if energy did not come in a continuous flow, as suggested by accepted electromagnetic theories, but instead came in separate, tiny packets or quanta. The singular, quantum, from the Latin for amount, is a discrete unit, the smallest possible packet that cannot be divided any further.

Planck's announcement of this theory in 1900 marked the birth of quantum theory or quantum physics, a profoundly new way of looking at the underlying principles of reality. Quantum theory would encompass concepts such as the observer-related universe, where the very act of observing an experiment affects its outcome; wave-particle duality, where a subatomic particle also behaves as if it is a waveform; and whether or not photons (light particles) communicate with each other. Although some of the ideas sound crazy, they only apply at the quantum level, not at the level of ordinary life.

In order to describe his theory in mathematical terms, Planck developed the equation that the energy of a vibrating molecule, E (measured in joules), is equal to its frequency, v (measured in hertz), multiplied by a new constant value, h (measured in joule-seconds): $E = hv$. This new constant he identified came to be called the Planck constant.

Radioactivity: Marie Curie

In 1896 the French physicist Henri Becquerel (1852–1908) first reported the 'unique activities' of uranium when he discovered radioactive waves emitted by this element. These waves acted somewhat like X-rays in that they could penetrate matter and ionize air.

Searching for a suitable research topic, Marie Curie decided to find out whether other substances also had 'unique activities'. With her husband Pierre she tested waste ore from which the uranium had been removed, and when this 'pitchblende' was still found to emit the odd rays, she realized that since uranium did not have to be present, there could be other substances with these activities, which they called radioactive. They discovered the new elements of polonium (named after her homeland, Poland) and radium. They also observed that the 'unique activity' was a chemical property affecting organic tissue. They had no idea that the damage to tissue from their new radioactive materials could be lethal.

However, the discovery of radioactivity inspired other scientists to do further research on atoms and the atomic structure. In 1901 Ernest Rutherford (1871–1937) helped discover the mechanism behind radioactivity when he found that atoms of some unstable elements would collapse into atoms of different elements, emitting charged particles in the process.

From 1915 Marie Curie began to train doctors in the medical use of radium to treat arthritis, scars and some cancers. Her research into therapeutic uses of radioactive material led to the development of medical X-rays, and during the First World War she helped to bring mobile radiography units to the field to help find shrapnel in wounded soldiers. They nicknamed these boxes '*petites Curies*' or little Curies.

Marie Curie (1867–1934)

Born in Warsaw, then part of the Russian Empire, Marie Curie studied at the University of Paris (the Sorbonne), graduating first in her class in 1893. In 1903 she became the first female Nobel laureate when she was awarded the Nobel Prize in Physics for her research into radiation, along with her husband Pierre and Henri Becquerel. She had many 'firsts': in 1909 she was appointed the first female professor at the Sorbonne. This was tempered by sadness since she was stepping into her dead husband's shoes. In 1911 she received a second Nobel Prize, in Chemistry, for her discovery of radium and polonium. This made her the first scientist to win Nobel Prizes in different disciplines.

Despite all her achievements, Marie Curie faced prejudice from male scientists and was not elected to the French Academy of Sciences. She died from leukaemia, assumed to be due to her exposure to radioactive material, and her notebooks remain radioactive today.

The Relatively Miraculous Year: Albert Einstein

The year 1905 became known as Einstein's 'miracle year' since he published four papers, each of which provided a major contribution towards an understanding of the universe.

First was his quantum theory of light. Max Planck (*see* p. 94) had previously suggested that energy is emitted in tiny units called quanta, and Einstein theorized that light is made up of quanta. These elementary light quanta are now called photons.

Secondly, he explained Brownian motion – the seemingly random movement of microscopic particles – as the movement of atoms. He shared this breakthrough with Marian Smoluchowski (1872–1917).

Third was his special theory of relativity. His later general theory of relativity (*see* p. 41) changed Newton's ideas of the gravitational field, and his special theory overthrew classical notions of absolute space and time. His idea was that time and space are relative to the observer: they can be perceived differently. For example, an atomic clock travelling in a jet plane ticks more slowly than a similar clock stationary on the ground because it is in a different state of motion.

Einstein's fourth paper that year, stating that energy and mass are interchangeable, provided the equation $E = mc^2$. Here he posited that a body's energy E equals its mass m multiplied by the speed of light c squared. The speed of light is so extreme that the conversion of even a tiny amount of mass releases a vast amount of energy.

Overall, Einstein's influence on science was incalculable. He overthrew Newtonian physics at the extremes of mass and speed, and introduced a whole new way of looking at the universe.

Albert Einstein (1879–1955)

Born into a German family of Jewish descent, Albert Einstein went to school in Switzerland and stayed there to work, gaining citizenship in 1901 when he obtained a clerical job at the patent office. The work was so intellectually undemanding that he was able to develop some of his scientific theories and pursue a doctorate. He later accepted posts at German scientific and academic institutions.

When the Nazis encouraged anti-Semitism in Germany, Einstein embarked on lecture tours abroad, and left Germany for good in 1932, settling in the USA and taking American citizenship in 1940. In 1952 he rejected the offer of the presidency of Israel.

A pacifist, one of his last acts was to call upon world leaders to use peaceful means to resolve conflicts.

Microwaves and Plant Physiology: Jagadis Chandra Bose

A pioneering Indian inventor, Jagadis Chandra Bose was ahead of his time in many ways since some of his ideas and discoveries were only accepted by the wider scientific community years after his death in 1937.

When Bose discovered the existence of very short wavelengths – a few millimetres in size – he called them 'millimetre waves'. In experimenting on them he incidentally built an improved radio detector, as well as several microwave components that are commonplace today – but it would be fifty years before other scientists made use of his discoveries of the quasi-optical properties of short radio waves.

Instead of just being neglected, his theories on plant physiology were actively refuted at the time. He created his own highly sensitive instruments to measure plant growth and reaction to external stimuli such as light, touch and temperature as well as deliberately unpleasant stimuli such as cuts or harmful chemicals. Bose was able to prove that plants' reactions to these were electrical in nature, not chemical as had been previously thought.

Further experiments showed that noise can have an effect on plant growth: plants grow faster and stronger when exposed to pleasant, soothing music, but their growth is retarded by harsh, discordant sounds.

Bose's findings were considered to be extraordinary at the time and many scientists would not accept that plants could behave in the same way as animals when stimulated. Nowadays this response of the nervous system is fully accepted, although it is considered to be a knee-jerk reaction. Not many people believe it shows that plants have full consciousness.

Jagadis Chandra Bose (1858–1937)

Sir Jagadis Chandra Bose was born in East Bengal (now Bangladesh). He attended a local village school before going to Calcutta (Kolkata) then Britain to finish his studies. Returning to India, he became the first native Indian to hold the post of professor at the Presidency College, Calcutta, but he was paid less than Europeans doing the same job. In protest, Bose refused to take his salary. He did such excellent work that eventually the university agreed to give him a pay rise and backpay.

Bose argued strongly that India needed to build a skilled, modern scientific base, and he was opposed to caste distinctions and to the religious conflicts between Hindus and Muslims.

Believing that knowledge should benefit all humanity, Bose initially did not file patents for his many discoveries. He was knighted for his services to science in 1917.

Nuclear Physics is Born: Ernest Rutherford and Niels Bohr

In 1904 J. J. Thomson (*see* p. 88) had proposed the 'plum pudding' model of an atom: that its very light, negatively charged electrons are spread evenly through the positively charged body mass so that the atom remains neutral overall. British physicist Ernest Rutherford (1871–1937) tested this by firing positively charged alpha particles (similar to helium and usually released by the decay of a larger atom) at a thin piece of gold foil. If the gold atoms were in balance, they would neither attract nor repel the particles, which would pass straight through to a detector screen. Eventually, he found this was not happening; some of the alpha particles were reflected off the foil.

In 1911 Rutherford proposed his own model: the atom was mostly empty space, with a tiny but dense positive nucleus, and electrons orbiting around it at the edge of the atom like planets in orbit.

Just two years later his model was superseded by that of Niels Bohr, who applied the principles of the new field of quantum physics, and thought that electrons could only occupy certain energy levels or orbitals around the nucleus depending on how much energy they carried. He calculated these levels by the centrifugal force of the electron's movement and the electromagnetic attraction between electron and nucleus, and checked them using spectral analysis.

It wasn't long before it was found that the nucleus was actually two particles: a positive proton and a neutral neutron. Then in 1964 it was discovered that these two are composed of even smaller particles, quarks, that come in six flavours: up, down, top, bottom, strange and charm. It's true – not even a quantum physicist could make it up.

Niels Bohr (1885–1962)

At school Niels Bohr's best subject was physical education; he only just missed out at representing his country, Denmark, at football.

In 1912 he began to work in England under Ernest Rutherford; then in 1921, after he developed his theory of atomic structure, a new Institute of Theoretical Physics was created in Copenhagen for him to head.

During the Second World War, Germany conquered Denmark. Bohr had some Jewish ancestry, so in 1943 he was included in the Danish Resistance's mass evacuation of nearly all the country's Jews to safety in Sweden. He was much sought after by the Allied scientists of the Manhattan Project, who were building an atomic bomb, so he was brought to Britain in the hastily converted bomb bay of a warplane. He passed out when he didn't put on his oxygen mask in time, and nearly died.

Bohr was one of only a handful of Nobel Prize winners whose children have also gone on to win the award: his son Aage received a Nobel Prize in Physics in 1975.

Bose-Einstein Statistics: Satyendra Nath Bose

In 1924 an unknown Indian lecturer, Satyendra Nath Bose (1894–1974), had his paper 'Planck's Law and the Hypothesis of Light Quanta' rejected by a science journal. He then took a brave step by sending it directly to Albert Einstein (*see* p. 97), who immediately had it published in a prestigious science publication. Suddenly Bose was an international science star.

Bose was proposing a new statistical approach to measuring subatomic particles and therefore a new way to derive the formula, originally created by Max Planck (*see* p. 94), for radiation of energy from a blackbody. Planck himself had used classical physics to derive his original formula, but Bose sidestepped this altogether. Instead, he used Einstein's approach which says that quanta or small energy parcels of light behave like particles (photons) as well as like energy waves. So, Bose said, let us treat the blackbody energy as if it is in a different state, a cloud of photons similar to the particles in any old cloud of gas. But, instead of treating each particle as if it is statistically independent, he suggested they should be statistically analysed as groups of particles within defined spaces that he called cells.

Subsequently known as Bose–Einstein statistics, this approach worked and was an important contribution to the emerging science of quantum statistics. It only applies to subatomic particles that can exist in the same quantum or energy state at the same time within an atom, and can therefore cluster in groups. These types of particles are called bosons in honour of Bose's original work on their behaviour, and they include light photons. Particles that cannot share the same quantum state are called fermions, and their behaviour is described by a different set of statistics.

Matrix Mechanics and the Uncertainty Principle: Werner Heisenberg

The German theoretical physicist Werner Heisenberg is best known for his uncertainty principle, outlined in 1927, which states that it is possible to determine a particle's position and momentum, but not both at the same time. It is therefore impossible to predict with accuracy a particle's path or position in the future. Like other aspects of quantum physics this applies only to very small particles such as atoms or parts of atoms.

The Heisenberg uncertainty principle contributed to theories of a non-causal or indeterministic universe in which, at the subatomic level, science can only suggest probabilities, not certainties. This aspect of new physics says an act is only determined at the point where a scientist observes or measures it, which 'fixes' the probability of the act. This was at complete odds with the deterministic nature of the universe held in classic, Newtonian physics.

Although some scientists opposed this whole idea, it was included as part of the paradigm of quantum physics adopted in 1927 as the Copenhagen Interpretation.

Even before proposing the uncertainty principle, Heisenberg had stamped his mark on quantum physics with the invention of matrix mechanics, the first mathematical formulation of quantum mechanics. He had been trying to discover a mathematical calculation to explain the spectral lines or light frequencies given off by the movement of a particle within an atom. His resulting formula represented the momentum and position of the particle as a matrix of coefficients indexed by the start and end energy levels. It used the standard maths of matrices, such as arrays of numbers, which could then generate an equation.

Werner Heisenberg (1901–76)

Like some other great scientists, as a young student Heisenberg excelled in mathematics and theoretical physics, but had a poor grasp of applied physics. He nearly failed to gain his doctorate because he could not explain how a battery works.

A native German, Heisenberg fell foul of the Nazis because of his work on quantum physics, which the Nazis saw as 'Jewish science' rather than the 'Aryan science' of which they approved. His appointment as professor at the University of Munich was blocked, but later on, during the Second World War, he was brought into the Nazi's atomic research group. He was the one German scientist the Allies feared might create an atomic bomb, so he was on their hit list.

It is not clear whether Heisenberg deliberately misled the Nazis when he announced that a bomb would be impossible to create. After the war he worked for peaceful uses of atomic energy and was one of the founders of CERN, the pan-European nuclear research council.

Wave Mechanics and Schrödinger's Cat: Erwin Schrödinger

In 1925 the French physicist Louis de Broglie (1892–1987) proposed that all subatomic particles might also have properties of waves. A few weeks later the Austrian Erwin Schrödinger invented wave mechanics as a way of mathematically describing this strange behaviour of particles. His approach was to treat the particles as three-dimensional waves, each with their own unique wave function, all governed by a fundamental differential equation, known as the Schrödinger equation.

Schrödinger always said that his differential equation was mathematically equivalent to Werner Heisenberg's algebraic approach to quantum mechanics (*see* p. 104), a statement that was later proven.

To illustrate the quantum uncertainty principle proposed by Heisenberg, and to show how bizarre quantum science was becoming, Schrödinger put forward probably the most famous thought experiment in history, 'Schrödinger's cat'. This imagines a cat in a sealed box with a tiny amount of radioactive material that in one hour has an even chance of decaying and releasing an atom. If it does decay, the quantum event will trigger a device to kill the cat. But until you open the box you do not know if the cat is alive or dead. Until then, Schrödinger said, the cat exists in two universes, one where it is alive, one where it is dead, and only by opening the box does the wave function collapse into an actual state of being.

Erwin Schrödinger (1887–1961)

Although born in Vienna, Austria, Schrödinger's mother was English so he grew up bilingual, speaking both English and German. His first academic posts were in experimental physics, which he said gave him a good grounding, and he was thirty-nine – unusually late for theoretical physicists – before he published his first paper on the theory of wave mechanics.

But he disliked the probabilistic nature of the new physics that he had helped to create, and after 1926 he worked in biology and on the still unsolved problem of a unified field theory. Like some other quantum physicists, he was drawn to Eastern philosophies.

Schrödinger was a well-known womanizer, and he had an open marriage. Both he and his wife had affairs and he had children with other women, scandalizing academic institutions around the world.

The Atom Bomb: Leo Szilard and Enrico Fermi

In 1939 the Hungarian-Jewish exile Leo Szilard (1898–1964) persuaded Albert Einstein (*see* p. 98) to write to President Franklin D. Roosevelt urging that the US should immediately begin work on an atomic bomb. Szilard had lived under the Nazis and was sure they would not wait to try to develop such a weapon. Eventually the USA, the UK and Canada launched the secret Manhattan Project to develop nuclear weapons.

Many great physicists living in Allied lands during the Second World War were drafted onto the project. They included New Mexico laboratory director J. Robert Oppenheimer (1904–67) and the team's youngest member, the brilliant science communicator Richard Feynman (1918–88).

Szilard had first considered the possibility of sustained nuclear power in the mid-1930s, and thought that a neutron chain reaction would be possible, whereby the nucleus of a certain type of atom would be forced to decay or release a neutron particle, causing a self-sustaining chain of decays, each releasing energy. Szilard then switched to nuclear fission, which involves bombarding a uranium atom with one neutron to cause the chain reaction.

He joined the Manhattan Project in 1942, working initially at the University of Chicago with Enrico Fermi (1901–54), who was an expert on radioactivity and had confirmed the existence of the neutrino or little neutron. He had also previously worked on bombarding atoms with neutrons. Other researchers such as Otto Hahn (1879–1968) and Lise Meitner (1878–1968) had already achieved nuclear fission.

In 1942 Fermi and Szilard completed the world's first atomic pile or nuclear reactor, and witnessed the first nuclear chain reaction. In their laboratory they carefully controlled the experiment, but in the bomb the chain reaction was allowed to continue unchecked.

Both men later opposed the development of a hydrogen bomb, with Szilard, who pleaded with the US military not to drop the atom bomb after all, becoming an active campaigner against the arms race.

The Standard Model and the Higgs Boson: Peter Higgs

There have been plenty of advances since the Second World War: harnessing nuclear energy for peaceful uses;

collecting solar energy; breaking the sound barrier; lasers, superconductors, transistors... Yet years into the twenty-first century it is clear that however many quantum particles there might be, Newton's laws of motion and gravitation still describe the physics of our everyday lives.

In 1964 British physicist Peter Higgs (born 1929) and other researchers proposed the existence of a boson particle responsible for moving mass to matter: the Higgs boson. Because of the wave-particle duality at the quantum level, the Higgs particle would be associated with a quantum field, which, it is theorized, would have been created by these bosons at the very moment of the birth of the universe, when energy was first converted to matter. The Higgs field would explain several anomalies in the behaviour of some subatomic particles but would also explain why particles have mass at all.

Since 2011 experiments have been carried out using the Large Hadron Collider, the largest particle accelerator in the world, at the CERN laboratories in Switzerland to try to identify the Higgs field or Higgs boson, and in 2012 CERN cautiously announced that they had identified a particle that behaved the way they predicted the Higgs boson would do. The existence of the Higgs boson has since been confirmed.

Perhaps similar evidence is out there for unified field theory, the boson particle of gravitation and even the alchemists' stone, all awaiting a future physicist to come along and investigate.

CHAPTER 4

Chemistry: Discovering Elements and Compounds

CHEMISTRY IS CONCERNED with the building blocks of the universe: the chemical elements, substances that cannot be reduced to any other matter. The element oxygen contains nothing but oxygen. Iron contains nothing but iron. But as chemical compounds, oxygen and hydrogen create water, and oxygen and iron can form rust.

Chemistry explores the basic substances of matter, investigates their individual properties and reactions to physical stimuli, describes how they bond together, and attempts to create new substances.

The story of chemistry covers ancient Greek philosophers, medieval magicians, exploding laboratories, tiny atoms and quantum mechanics. One of its most important tales is the creation of the periodic table, a way of organizing the elements that lets chemists the world over know an element's properties at a glance. From the four elements that satisfied the ancients, the periodic table currently lists 118, but chemists predict the discovery of yet more elements as we investigate the basic matter of the universe.

Ancient Elements, Early Science and Alchemists: Hypatia

For a large chunk of history, humanity made do with just a handful of material elements. There were five elements in some ancient cultures; for example, Babylon had wind, fire, earth, sea, sky; and China had earth, fire, water, metal, wood. But the classical four – earth, air, fire and water – were all that were recognized in the West for centuries, although the Greek Aristotle (*see* p. 16) added a fifth, unchanging, heavenly element, aether.

Around the third century BCE, the centre of learning in the classical world began to shift from Athens to Alexandria in Egypt. Living in this city hundreds of years later was one of the world's first female scientists, Hypatia. As a scholar, she began exploring the properties of liquids. She may have discovered that elements can take different forms and yet remain the same elements, for example, water can be frozen into ice and iron can be heated until molten, but it would be a long time before scientists understood that it is the changing arrangement of an element's molecules that determines its physical form.

Hypatia's knowledge was restricted to more visible properties, but she is credited with inventing a hydrometer to measure the relative density and gravity of liquids.

While Hypatia was observing, testing and inventing, others in Alexandria were doing the same thing for different reasons. Alchemy is thought to have begun in the city around the fourth century CE, as magic and the occult began to permeate. Alchemists mainly sought magical secrets or to transmute base metals into gold.

The word 'chemistry' comes from 'alchemy', which probably derives from the ancient name for Egypt: Khem. The alchemists of the Middle East and Europe unwittingly helped develop chemical knowledge (for example, of metals) and techniques, such as Alexandrian alchemist Mary the Jewess' bain-marie, still used today to gently heat 'vital' substances like chocolate or caramel.

Hypatia (*c.* 350/370 to 415 CE)

The daughter of a mathematician who was the last head of the great Library of Alexandria, Hypatia was a Greco-Egyptian. She was educated in Athens before returning to her birthplace of Alexandria, which was then part of the Byzantine Empire, and around 400 CE became a teacher of philosophy and astronomy. She was a leading Neoplatonist and one of the last classical scholars.

Unusually for her time, Hypatia refused to wear traditional women's clothes, and instead wore the robes of a scholar.

According to contemporary sources, Hypatia was murdered by a Christian mob after being blamed for a local conflict.

Birth of Chemistry as a Science: Robert Boyle

Alchemy was not confined to the Middle Ages; the seventeenth-century scientist Robert Boyle (1627–91) was primarily an alchemist. Yet he is credited with being the first to make a distinction between alchemy as an occult enterprise and chemistry as a science.

He did this with the publication in 1661 of a landmark book, *The Sceptical Chymist*, in which he laid out his own systematic approach to experiments and condemned the superstitions, contradictions, erratic beliefs and actions of most alchemists.

The son of an Irish earl, Boyle did most of his experiments with gases. At the time, air was held to be a single substance, not a mixture of gases. On hearing of the recent invention of an air pump by the German Otto von Guericke (1602–86), Boyle built his own improved version so he could create vacuums or control the amount of air in a vessel. He proved that air is necessary for life and for flame, that sound does not carry in a vacuum, and that air is permanently elastic. This led to the formulation of Boyle's law on gases, which states that the volume occupied by a gas is inversely proportional to the pressure of that gas.

Boyle was born six years after the English lawyer Sir Francis Bacon (1561–1626) published his proposal for a scientific method. Like others at the time, the Irishman began to query the classic view that there were only four elements. Chemistry was on the cusp of a new scientific beginning, but Boyle would never lose his belief that it was possible to transmute base metals into gold.

The Chemical Revolution: Antoine-Laurent Lavoisier

By the mid-eighteenth century new substances had been discovered and named; for example phosphorus, found through a convoluted process of treating urine. Through more salubrious means, carbon dioxide ('fixed air'), hydrogen (speculated to be 'phlogiston', the essence of fire), nitrogen (known as 'phlogisticated air') and new metals (actually elements) such as barium, molybdenum and tungsten were found. Scientists were also approaching a greater understanding of compounds, or how substances combine.

In 1789 the French chemist Antoine-Laurent Lavoisier published a table of elements, finally listing more than the classical four, even though some of his thirty-three were wrong. Working with other scientists, Lavoisier also developed a new naming system for chemicals, which is fundamentally the one used now to reflect substances' known compositions.

A wealthy man, Lavoisier had a beautifully equipped laboratory where, although he seldom originated work, he was able to confirm and attempt to explain others' ideas, sometimes causing arguments about precedence, particularly with British chemist Joseph Priestley (1733–1804) over the discovery of oxygen.

Lavoisier's particular achievement was to demonstrate the role of oxygen in combustion, laying to bed the theory that 'phlogiston' or heat in the air was responsible. Through careful measurements in sealed containers, Lavoisier showed that 'dephlogisticated air', as Priestley called it, was the substance burnt up in air, and was the same substance that was absorbed from air by metal residues after heating. He called this gas 'oxygen' meaning 'acid generator' since he mistakenly thought that is what it also did.

So Lavoisier named the air that we breathe, and his standardization of chemical descriptions and methodology meant that he helped revolutionize chemistry.

Antoine-Laurent Lavoisier (1743–94)

A French aristocrat, Lavoisier welcomed the rational policies of the new regime that came into place following the French Revolution of 1789. He stayed in Paris to become head of the Academy of Sciences and to continue his work heading the National Gunpowder Administration, ensuring that future revolutionary armies would be self-sufficient in gunpowder.

However, before the revolution Lavoisier had been a 'tax farmer' who invested in the finance company General Farm, which lent money to the government and collected some taxes as reimbursement. Tax farmers usually became wealthy and were deeply unpopular.

When the Reign of Terror began to threaten everyone who had benefited from the old regime, Lavoisier continued to believe that as a useful and loyal scientist he would be safe. He was wrong and he, along with many other tax farmers, was arrested. His scientific work was cut short by the guillotine.

Electrochemistry: Humphry Davy

While studying science at Bristol in England, Humphry Davy (1778–1829) did what many students do today and experimented with inhaling nitrous oxide or laughing gas. His report on his gaseous experiences won him a position as assistant lecturer at the Royal Institution in London, where he became popular with fashionable society.

Electrical science was in its infancy, and Volta's electric 'pile' (*see* p. 86) had only just been invented. In 1807 Davy and other members of the Royal Institution built their own electric battery out of silver-zinc cells, the most powerful one in the world at the time. Davy knew that electricity could split compounds (electrolysis), so directed the electrical power at the 'earths' of caustic potash and soda. Antoine-Laurent Lavoisier (*see* p. 116) had included these in his list of elements, but Davy was able to prove they were compounds, since they broke down into potassium and sodium respectively.

Davy continued his experiments and discovered magnesium, calcium, boron and barium. He also proved that chlorine gas was an element and would not break down any further, as some scientists had predicted.

But to the general public Davy is known as the inventor of the safety lamp for mining. That perilous occupation involved carrying naked flames into the mines where flammable gases would often escape from the workings. Davy's lamp had a metal mesh surrounding the flame, preventing it from escaping to ignite gases outside the lamp.

Atomic Theory: John Dalton

Back in ancient Greece a relatively unpopular theory about the structure of matter had been proposed by Leucippus and Democritus that the universe is made of tiny, indivisible solids called atoms, derived from the Greek for 'uncuttable'. Over in India, Buddhist thought also held that matter was made up of minute, basic units. But in the West it was Aristotle's view of the four earthly elements (see p. 16) comprising matter that held sway for thousands of years until an English meteorologist explored the fundamental nature of gases.

John Dalton (1766–1844) could not recall exactly what led him to the theory that gases – and all elements – are made up of tiny atoms (adopting the Greek word) that are unique to each element and are distinguishable by their relative weights. He came to the conclusion that chemical compounds are formed when the atoms from two different elements are combined, and chemical reactions occur when the atoms are rearranged. Finally, he theorized that atoms cannot be manufactured or destroyed.

Dalton believed he could prove his theory since a container of hydrogen weighed less than a container of oxygen, so their basic units must be different. Also, even when mixed in a container, different gases diffused (spread out) as separate entities, so must be composed of unique units. Defining hydrogen, the lightest gas, as 1, he attempted to find the atomic weights of other gases relative to this by seeing how they combined with a set mass of hydrogen.

Dalton proposed that atoms cluster together into molecules (a recently coined word for a very small particle). He also proposed that elements combine in fixed ratios, such as carbon and oxygen in the ratio of 1:1 to form carbon monoxide, and 1:2 for carbon dioxide. He got one thing wrong, though: he thought water was hydrogen and oxygen in the ratio of 1:1. Frenchman Joseph-Louis Gay-Lussac (1778–1850) corrected this to 2:1, opening the floodgates to chemical explorations of oxygen compounds.

Although atomic theory is nowadays associated with particle physics, whereas Dalton's work was in chemistry, he showed how the two disciplines are linked together, forming a basis for many modern integrations such as electrochemistry, radioactivity, nuclear physics and quantum chemistry.

Isomerism and Organic Chemistry: Justus von Liebig

Could substances with similar molecular formulas, that is similar combinations of atoms, actually behave differently? Studying independently, in 1827 the German chemists Justus von Liebig and Friedrich Wöhler (1800–82) found that they could – if their molecules are arranged in different structures. Even small changes of shape on the molecular level can result in changes in their properties on the larger scale. They had discovered isomers, a term coined by Jöns Jakob Berzelius (1779–1848) in 1830. He was a Swedish chemist who, incidentally, also introduced the Latin notation for chemical symbols (for example, Fe for iron from the Latin *ferrum*).

Isomers can have different bonds within the molecule, or can be mirror images of each other. Nowadays they are particularly common in medicinal chemistry. For example, phentermine is an appetite-reducing drug, but rearrange its atoms and we get the strong stimulant dextromethamphetamine.

Von Liebig's initial interest in organic chemistry, the study of carbon-containing molecules, changed to applied chemistry, particularly the application of chemical science to food, agriculture and nutrition. In 1838 he wrote: 'The production of all organic substances no longer belongs just to the organism. It must be viewed as not only probable but as certain that we shall produce them in our laboratories.'

He followed this up by producing a cheap extract of meat. He also explored, through soil analysis, the best way to feed crops and produced his own manures. In the process he proved that a plant's carbon content comes not from leaf-mould or humus, but from photosynthesis.

In 1908 another German, Fritz Haber (1868–1934), invented a mass process to take nitrogen from the air and turn it into life-giving fertilizers. He went on to invent chemical weapons for use in the First World War.

Justus von Liebig (1803–73)

Liebig was born in Darmstadt, Germany, where his father, a chemical manufacturer, owned a shop and, more importantly for his son, a laboratory where the boy could experiment with chemicals. The young Liebig's experiments not only brought about an explosion at his school, but also caused structural damage at his home. His parents' decision to apprentice him to an apothecary may have been as much to keep their house in one piece as to support the boy's career.

Liebig became a professor at the University of Giessen when he was only twenty-one. He was a radical teacher, insisting that chemistry should be an independent subject, not just part of pharmaceutical studies. Perhaps remembering his own mishaps, he also instigated the practice of controlled laboratory experimentation. His methods became the standard model around the world.

The Law of Substitution: Jean-Baptiste Dumas

In the early nineteenth century the accepted theory of molecular structure was that all chemical compounds were either positive or negative, and chemical combinations resulted from the attraction of the opposing parts. This 'dualistic' theory was particularly championed by Jöns Jacob Berzelius.

But the French teacher and politician Jean-Baptiste Dumas (1800–84) found that burning candles bleached with chlorine produced fumes of hydrogen chloride, and concluded that 'during the bleaching, the hydrogen in the hydrocarbon oil of turpentine became replaced by chlorine'. He had shown that in certain circumstances it was possible to substitute the hydrogen atoms (that were electropositive) with chlorine or oxygen atoms (electronegative) without there being any dramatic structural change.

Berzelius and others, including Justus von Liebig (*see* p. 121), disputed these findings so strongly that Dumas retreated from his position. But in the long-run this theory did supersede that of Berzelius.

Less contentious was his work identifying the components of compounds such as urethane and methanol by distilling wood. He also improved methods of measuring vapour densities by finding the mass, temperature, volume and pressure of a substance in its vapour phase. This led directly to the more accurate atomic weights of thirty elements, half the total number known at the time.

Dumas was a pioneer in the study of organic chemistry. Like Liebig, he was also one of the first teachers to insist on scientific laboratories and rigid experimentation. But, despite all his achievements, he was not above using his eminent position in the Academy of Sciences to hinder the careers of younger chemists whom he feared threatened his reputation.

The Bunsen Burner: Robert Wilhelm Bunsen

Although the German Robert Wilhelm Bunsen (1811–99) made a number of discoveries and inventions, he is primarily known for the gas burner that bears his name. He produced this in 1855, based on one developed by Michael Faraday (*see* p. 89), and it has transformed the practice of chemistry.

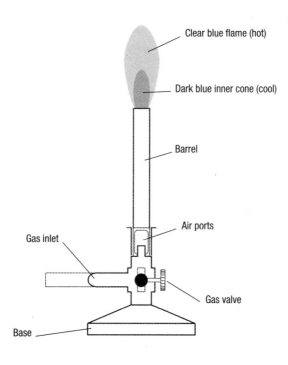

The Bunsen burner.

He was inspired by his poorly built university laboratory, which was supplied with potentially toxic coal gas. Bunsen needed a reliable gas offering both light and heat, and designed a burner with air holes near the bottom, allowing the gas and air to mix before ignition and giving rise to a tall flame. The air flow could be increased to produce a clean, hot, blue flame that was suitable for use with laboratory glassware.

The blue of Bunsen burners did not interfere greatly with the colours of burning elements, so contributed to the emerging science of spectroscopy, the study of the colours or spectrum of light emitted from flames of different elements. Bunsen and Gustav Kirchhoff (1824–87) used such analysis to identify a new alkali metal, caesium.

The Periodic Table: Dmitri Mendeleev

The periodic table sits at the heart of chemistry. Listing all the elements in a way that shows their basic properties and the groups to which they belong, it is a diagram that can be instantly read by chemists the world over.

Russian Dmitri Mendeleev was not the first person to attempt to catalogue the known elements. Others, particularly English chemist John Newlands (1837–98), had noticed patterns in the chemical properties of elements. Newlands proposed a 'law of octaves', since it appeared that the properties seemed to fall into seven groupings, with every eighth element starting a new row.

Mendeleev's basic listing is in order of atomic number – the number of protons on the atom's nucleus. He added valence as another variable. This is the combining power of an atom, relating to how many electrons are in its outer shell and therefore are available for combining, and corresponding roughly to its atomic number. Elements with the same valences line up underneath each other on the table in a periodic pattern, forming vertical groupings. For example, the noble gases end up in a column underneath each other, as do the halogens. So Mendeleev's periods are the horizontal rows of the table, while the vertical columns are known groups with similar properties. Just as important, he argued convincingly – and correctly – were the gaps in his table, which represented as yet unknown elements.

Mendeleev's table worked, even if no one at the time was absolutely sure why. As it happens, the periodic patterns that he had noticed are explained today by the number of electrons on an atom's outermost shell.

Dmitri Mendeleev (1834–1907)

Born in a small town in Russian Siberia, Mendeleev was the youngest of the fourteen surviving children of a teacher, although the exact number of children varies in different sources. He went to St Petersburg at the age of fourteen to continue his education, and spent much of his life there, obtaining his first professorship in 1864.

Gaining a fellowship, he studied abroad for two years at the University of Heidelberg, but instead of forging close links to the university's other chemists, including Robert Bunsen (see p. 123), Mendeleev set up his own laboratory in his apartment.

In 1860 he attended the International Chemistry Conference held in Karlsruhe, Germany, where he came into contact with new ideas on atomic weights, information that helped him clarify how to organize the elements.

Statistical Mechanics and Thermodynamics: Josiah Willard Gibbs

In the early 1870s physical chemistry consisted of just isolated observations and facts. It is an area that studies chemical systems through the eyes of the laws of physics, applying concepts and principles such as energy, force and motion. This includes exploring the rate of chemical reactions through kinetics (the study of motion and its causes), or looking at the forces that act upon materials, affecting their tensile strength or plasticity.

The discipline began to change in 1875–8 with the publication of a 300-page paper *On the Equilibrium of Heterogeneous Substances* by American mathematician and physicist Josiah Willard Gibbs (1839–1903). Containing 700 mathematical equations, the book brought together and explained discoveries in physical chemistry, but also proposed Gibbs' own ideas. He felt that thermodynamics, the study of the relationships between heat and temperature, and energy and work, could be helpful in exploring and explaining chemical states.

A theoretician at heart, Gibbs, together with James Clerk Maxwell (*see* p. 91) and Ludwig Boltzmann (1844–1906), developed statistical mechanics, a term he coined to explain the thermodynamics as a consequence of the statistical properties of large assemblies of particles. Gibbs also used statistical mechanics to analyse chemical substances and reactions.

Gibbs introduced the concept of chemical potential, which is the rate of increase of the internal energy of a system according to the increase in the number of its molecules, and he described free energy, a measurement of the thermodynamic state.

His work was difficult to understand, even for other theoreticians. When James Clerk Maxwell died prematurely in 1879, a joke went around in America that there had only been one man who could understand Gibbs, and now he was dead.

Biochemistry and Synthesizing Compounds: Emil Fischer

The German organic chemist Emil Fischer (1852–1919) made many important discoveries of the structures of sugars and proteins, and of the nature of purines (certain compounds sharing a chemical base). His descriptions of carbohydrates and amino acids helped form the discipline of biochemistry.

Fischer's work on purines lasted seventeen years, beginning in 1882 when he proved that several seemingly unrelated natural compounds were chemically linked. Some were animal products such as uric acid, and others, including caffeine and theobromine (found in chocolate), were from plants, but they all shared a common atomic group of 'five carbon atoms and four nitrogen atoms so arranged that two cyclic groups with two common atoms are formed'. He named this common link 'purine' and found that all purines could be derived from each other.

Fischer was always keen to synthesize compounds, which not only proved their structure, but would, he hoped, supply cheap medical or even food products. Barbiturate drugs are one result of his work. He managed to synthesize several purines and, after investigating sugars, also managed to create synthetic glucose and fructose. Fischer discovered that enzyme activity is determined by molecular structure, not content. He realized this when he found that yeast enzymes (natural proteins that affect chemical changes but are not altered themselves) only eat sugar isomers of certain shapes.

Fischer also synthesized several amino acids, and discovered the chain linking amino acids together. All in all, Fischer's body of work greatly advanced physiological research. Unfortunately, his early doctoral research on phenylhydrazine probably gave him cancer, and he also suffered from mercury poisoning. Early chemistry was dangerous work.

The Noble Gases: William Ramsay

By the nineteenth century, elements – from gases to metals – were being rapidly discovered and their chemical niches settled. But in the 1890s a group of gases was isolated that did not appear to interact chemically with other substances.

Since these gases had zero valence, there also seemed to be no place for them on the periodic table. They were too 'noble' to play with ordinary elements.

The first samples of these gases were collected by Scotsman William Ramsay (1852–1916) in the 1890s. Together with Englishman Lord Rayleigh, he first found argon when they realized the density of nitrogen collected from the atmosphere did not match that of nitrogen produced by chemical reactions. Isolating out all known gases, they found a minute amount of a new one they called argon, after the Greek for 'lazy', since it didn't seem to do anything.

In 1898 Ramsay found even rarer gases of the same family – neon, krypton and xenon – by first liquefying air, then heating it up and collecting each of its gases as they boiled away. Together with helium, which Ramsay had found in 1895, and radon, identified in 1900, these gases are chemically inert – they do not react with other elements. But a place was found for them on the periodic table as an eighth group.

The noble gases can produce bright, glowing colours under certain circumstances, meaning that rare neon is now better known than many common elements.

Free Energy and Covalent Bonds: Gilbert N. Lewis

Twenty years after Josiah Willard Gibbs (*see* p. 126) had formulated chemical thermodynamic theory, there had been little practical work in physical chemistry (possibly because Gibbs was so difficult to understand!).

The American Gilbert N. Lewis (1875–1946) tried to fill the gap by measuring the unknown free-energy values of chemical reactants, which vary according to the thermodynamic state. He also experimented on how to measure the entropy, or unavailable energy, of a system. His work helped predict whether chemical reactions would not go ahead, would go to completion or would reach equilibrium.

Lewis made another important contribution by theorizing that elements bond according to an atom's valence (the number of electrons in its outermost shell). He first developed this idea in 1902, when physicists were only just realizing that electrons were arranged in particular orders around the nucleus. He imagined an atom as a cube with a space for electrons on all corners, and proposed that chemical bonds are formed when atoms exchange electrons so that each has the ideal arrangement of filled corners. He continued to refine his ideas and in 1916 proposed that the chemical bond occurs when atoms actually share electrons. This discovery was later named covalent bonding. When an electron is not shared Lewis described it as a 'free molecule', but nowadays we call them 'free radicals' and chase them out of our bodies with antioxidants.

Chemical Bonds and Protein Structures: Linus Pauling

The American Linus Pauling, one of the most important chemists of the twentieth century, first began exploring the nature of the chemical bond in the 1920s, often using the ideas of quantum mechanics. Like many modern chemists, he applied techniques that had been developed by physics. Among his groundbreaking discoveries was the fact that within molecules of compounds the electron orbital paths (*see* p. 102) are sometimes hybridized, or combined. He also showed that ionic bonding, where electrons are exchanged between atoms, is an extreme case, as is covalent bonding, where electrons are shared. Later, in 1949, he and his research team identified the molecular basis of sickle-cell anaemia (*see* p. 189).

In the 1950s he turned his attention to identifying the structure of protein molecules. These are large, fragile and complex, so Pauling used his own model-building approach, first learning about the structures of the building blocks of the molecules – in this case amino acids – then looking at how they are linked, and finally building a model to test his findings.

Again using new techniques, he analysed protein molecules through X-ray diffraction, in which the substance scatters an applied X-ray and the diffraction pattern provides information about the lattice of atoms.

Pauling and his team developed a theory that amino acids are bonded together only at their ends, forming a rigid structure, and he was able to identify the three-dimensional helical structure – the alpha helix – which is a component of most proteins.

He was less successful in producing an accurate model for DNA, since he again proposed a triple helix – one helix too many.

Linus Pauling (1901–94)

Born in Portland, Oregon, USA, Pauling became fascinated with chemistry when he was fourteen and witnessed impressive chemical reactions from a friend's chemistry set. He straight away set up his own laboratory in the basement of his home, and in 1917 he began a course in chemical engineering at Oregon Agricultural College (now Oregon State University). Before long, while still an undergraduate, he was asked to teach the subject to other undergraduates, such was his advanced knowledge of the field.

In the 1930s Pauling began to focus on the structure of large biomolecules (those present in the living body), and in 1954 he was awarded the Nobel Prize in Chemistry. From the 1960s he began to spend more time on peace activism, calling for the scientific world to ban nuclear bomb tests, and he won the Nobel Peace Prize in 1963, thereby becoming the only person to win two unshared Nobel prizes.

Synthetic Chemistry: Elias James Corey

Building on the work of Emil Fischer (*see* p. 127) and others, organic synthesis is the production, via chemical processes, of complicated organic compounds using simple starting materials.

Synthetic chemistry has produced items as varied as nylon, plastics, paints, pesticides and many pharmaceutical products.

The traditional way of designing syntheses of complicated organic molecules (the target molecule) was to begin with simple or readily available materials and then keep trying to assemble them, by a sequence of chemical reactions, to form the target. Often chemists found it difficult to explain how exactly they came up with the starting materials or the process of the chain of reactions. In the 1960s the American Elias James Corey (born 1928), professor of chemistry at Harvard University, realized that a more planned and structured approach was needed, so he developed the principles of retrosynthetic analysis.

This technique involves a logical and systematic approach, starting with the target molecule and analysing how it can be broken down into smaller subunits. These are then also disassembled to end up eventually with simple starting materials. After working backwards in this way, it is then possible to build the target molecule, simply, quickly and efficiently. Every step is recorded and reversible.

With this widely applicable method Corey's research group was able to synthesize more than 100 products, particularly pharmaceutical substances. These included prostaglandins and other hormone-like substances that are used to induce labour, treat blood clots, allergies and infections, and control blood pressure. Some occur only in small quantities in nature, but thanks to retrosynthetic analysis, can now be found on the shelves of hospitals and pharmacies around the world.

Femtochemistry: Ahmed H. Zewail

Chemical reactions happen so quickly that they can only be described in tiny divisions of a second, known as femtoseconds. One femtosecond is just 0.000000000000001 of a second, or 10^{-15}. During the transition state of a chemical reaction the atoms of a molecule move extremely rapidly, taking less than 100 femtoseconds to rearrange themselves.

In the 1970s most scientists thought that it would never be possible actually to see what happens during such a fast reaction. But working at the California Institute of Technology, the Egyptian-born chemist Ahmed Zewail (1946–2016) realized that the recently developed technology of fast lasers might supply the sort of super-fast 'camera' that chemistry needed. Fast lasers are able to produce flashes of light that last just a few femtoseconds in duration, and in the 1980s Zewail began to experiment by using a series of flashes to initiate a chemical reaction and record the changes.

Zewail eventually developed a process involving mixing molecules in a vacuum tube and then beaming pulses at the mixture from a fast laser. The first flash of light energizes the chemicals, starting the reaction, then subsequent beams of light record the resulting light patterns or spectra from the molecules. These can then be analysed to see how the molecules are structurally changing.

To a chemist, this was the revolutionary equivalent of watching as the chemical bonds are broken and reformed. Instead of just imagining what was happening to atoms and molecules, they were now able to 'see' chemical reactions, and more easily plan experiments and predict results.

Zewail's technique became known as femtosecond spectroscopy or femtospectroscopy, and the whole body of work became the new field of physical chemistry called femtochemistry. It offers important applications in areas ranging from the development of medicines to the design of electronics. Zewail himself called it the ultimate achievement in the race against time.

The use of techniques from physics is clearly an important part of chemistry's future, from the possibility of building structures out of water to silicon nanotubes. The periodic table is not full up, and there are compounds yet to be synthesized. The alchemists of tomorrow will have plenty on which to work their magic.

CHAPTER 5

Biology: Characteristics of Life on Earth

LIFE ON EARTH is truly diverse, from the largest animals and plants down to microorganisms invisible to the naked eye, and yet all living things share certain processes. They all undergo the physical and chemical changes known as metabolism, which involves processing food to produce energy for growth and reproduction.

Ancient civilizations identified animals and plants that were useful to them, or which they feared, or were inedible, and they created the first classifications of living things. Herbalists developed knowledge of the medicinal uses of plants, and anatomists, through dissections and observations, began to discover how human and animal bodies work.

European explorers arriving in the New World in the fifteenth century found a diversity of species they had never seen before, and the microscope in the sixteenth century brought microorganisms and the basis of all life, the cell, into focus. Classification systems for plants and animals became ever more complex until the eighteenth-century Swedish scientist Carolus Linnaeus hit on a simple universal system that forms the basis of today's taxonomy.

Modern biology begins with Charles Darwin's theory of evolution by natural selection, tracing the rise and fall of biological traits in organisms. This led to genetics, and to cellular and molecular biology, which paved the way for human control of biological processes, such as genetic engineering and synthetic biology, to help industry and medicine. The proliferation of experimental research areas in biological science today goes to show just how far this discipline has come since the early studies in natural history.

Classification of Living Things: Aristotle

Although better known as a philosopher, the ancient Greek scholar Aristotle (*see* p. 17) was interested in every aspect of the natural world and has been described as the world's first great biologist. An early empiricist, he was a painstaking observer, amassing huge amounts of data on the behaviour and structures of animals and plants, and classifying more than 500 different species – work that was carried on by his student Theophrastus (*c*. 370 to *c*. 285 BCE).

Aristotle thought that each species had been designed for a particular purpose, and that species were fixed and unchanging, a view that would prevail until Darwin's theory of evolution was published in 1859. His unique contribution was a classification system for all living things. He grouped all known organisms under two main headings: plants and animals. Animals were divided into three groups according to where they lived: on land, in water or in the air. Vertebrates were distinguished from invertebrates and named, respectively, animals 'with blood' and 'without blood'. Animals 'with blood' were divided according to reproduction: live-bearing (mammals) and egg-bearing (birds and fish). Animals without blood were insects, crustacea and testacea (molluscs).

He introduced an early binominal system, giving two names to each organism: a generic or family name (the genus), and a second name to distinguish different members of the family by some unique characteristic.

Aristotle's system prevailed for 2,000 years and formed the basis for Linnaeus' classification in the eighteenth century, but it had flaws. For example, frogs are born with gills and live in water, and yet they grow up to have lungs, so they fall under two classes: 'living in water' and 'living on land'. His deductions were sometimes wrong, too, including his conclusion that flies are spontaneously generated from rotting manure.

Many Western scholars believed in the Great Chain of Being or Ladder of Nature, deriving from Aristotle and Plato's concepts of hierarchical classification. God came at the top of this scale of perfection, along with the angels, followed by kings and men, then animals, and plants and minerals came last, at the bottom. The hierarchical structure was believed to have been decreed by God and it held sway well into the Middle Ages.

Nature under the Microscope: Antonie van Leeuwenhoek

Science in seventeenth-century Europe was revolutionized when Antonie van Leeuwenhoek discovered the microscopic world.

Leeuwenhoek was a Dutch lens-maker who made more than 500 microscopes, and he was the world's first microbiologist. His microscopes were simply powerful magnifying glasses, not like modern microscopes with compound or multiple lenses, and yet he achieved 300 times magnification when other instrument-makers of the period could only magnify 30 times the natural size.

Leeuwenhoek's clear single lenses were held between two metal plates, riveted together and fixed three or four inches above the base of the instrument. He kept some of his techniques secret, but may have used the properties of spheres to improve his images, perhaps encasing his specimens in spherical drops of fluid.

Curious about everything that could be placed under a microscope, he examined plant and animal tissue, insects, fossils and crystals. As a result he was the first person to describe many microscopic aspects of life, such as living spermatozoa, and his discoveries disproved the commonly held belief that lower forms of life could be spontaneously generated out of the corruption of natural material, such as fleas from sand or dust, or flour mites from rotten wheat. Leeuwenhoek showed that these tiny creatures have the same life cycles as larger insects.

One of his important discoveries was unicellular, or one-celled, organisms. In 1674 he wrote about pond water: 'I found floating therein divers earthy particles, and some green streaks, spirally wound ... The whole circumference of each of these streaks was about the thickness of a hair of one's head.' He had discovered the single-cell structure of *Spirogyra* algae. The Royal Society thought his report so unlikely that it sent a special mission comprising a vicar, medical doctors and lawyers to examine his studies, but in 1680 Leeuwenhoek's observations were fully confirmed.

Leeuwenhoek also discovered bacteria, coining the word 'animalcule' (little animal) for many of the tiny creatures he saw: 'there were many very little living animalcules, very prettily a-moving. The biggest sort ... had a very strong and swift motion, and shot through the water (or spittle) like a pike does through the water.'

Antonie van Leeuwenhoek (1632–1723)

Seventeenth-century scientists were generally university-educated and from wealthy backgrounds, but Leeuwenhoek came from a family of tradesmen.

At sixteen, he was apprenticed to a textile merchant in Amsterdam, at a time when magnifying glasses were used to examine the quality of cloth by counting the density of threads in material. The glasses were fixed on a stand and could magnify up to the power of three. Thus Leeuwenhoek would have encountered the principle of magnification, but it is thought that his interest in exploring the microscopic natural world was sparked by a popular book *Micrographia*, written in 1665 by English scientist Robert Hooke (1635–1703). It contained reports and vivid pictures of tiny objects and creatures such as fleas and lice.

Within three years Leeuwenhoek was grinding lenses to make his own microscopes, and his observations would be reported over many years by the Royal Society of London – today the world's oldest surviving society for the advancement of science.

A Universal Classification: Carolus Linnaeus

Also swept up in the Scientific Revolution were botanists trying to name and classify all the new plants and animals discovered by explorers. The Swiss naturalist Konrad von Gesner (1516–65) grouped plants by their fruits; the Italian botanist Andrea Cesalpino (1519–1603) grouped by seed-bearing structures; and there were many more competing systems. But it was the Swedish-born Carolus Linnaeus (1707–78) who realized the importance of a universal taxonomy based upon observable characteristics and natural relationships.

Known as 'the little botanist' at eight years old, Linnaeus was inspired by his family's well-stocked garden. At the age of twenty-two he had collected more than 600 native plants, and he so impressed the eminent botanist, Olof Celsius (1670–1756), that Olof lent him a library and encouraged him to develop a new system of plant classification. Linnaeus brought back specimens from his travels and sent students on exploratory voyages, including Daniel Solander (1733–82), the Swedish naturalist who joined James Cook's world voyage in 1768 and returned with the first specimens from Australia and the South Pacific.

Linnaeus' major work, *Species Plantarum* (1753), named and described 7,300 species of plants using a binominal system (two names, genus and species, for each organism). All plants of one genus shared the same short Latinized name, where possible reflecting a characteristic of the plant, such as the genus *Helianthus* (meaning 'flower of the sun') for a group of flowers 'modelled on the sun's shape'. Up until then, plants had been given a group name followed by an unwieldy, long description, and names had often differed according to each botanist's system, causing much confusion. Linnaeus' simplification was revolutionary. For example, the wild dog-rose, previously named *Rosa sylvestris inodora seu canina*, became *Rosa canina* in Linnaeus' system.

Linnaeus' classification of nature into a hierarchy was equally successful. At the top were kingdoms; these were divided into classes, which in turn were subdivided into orders, and then into genera (genus is the singular), which were divided into species.

But not everyone was happy. Theologians criticized the classification of humans among primates, as it lowered the spiritually higher position of man in the Great Chain of Being (*see* p. 139). Linnaeus responded that 'animals have a soul and the difference is of nobility'.

Others were shocked by his classification of plants according to reproductive organs (stamens and pistils) and comparisons with human sexuality; for example, a plant with nine stamens and one pistil was likened to 'Nine men in the same bride's chamber, with one woman'.

Despite these setbacks, the Linnaean system was accepted. Today, although many new techniques for classifying and naming plants are used, based on plant genetics and biochemistry, Linnaeus' work lies at the core of modern taxonomy.

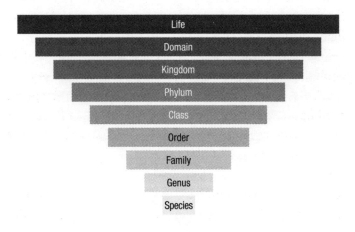

The fundamental principles of Linnaeus' classification of nature remain today, although in modern taxonomy the ranks of 'family', 'phylum' and 'domain' (or 'empire') have been added.

Building Blocks of Life: Matthias Schleiden, Theodor Schwann and Oscar Hertwig

Cells, the smallest structural and functional units of living organisms, were discovered in the seventeenth century after the invention of the microscope. For years scientists debated the exact nature of cells until finally, during an after-dinner discussion in 1838, two German scientists, Matthias Schleiden (1804–81) and Theodor Schwann (1810–82), came up with what is now known as cell theory. They deduced that the cell is the basic unit of life; that all living organisms are made up of one or more cells; and that all cells arise from pre-existing living cells – they are not spontaneously generated from non-living matter, as had once been thought.

Cell theory lies at the core of modern biology and is equivalent in importance to atomic theory in chemistry.

Schleiden was among the first scientists to recognize the importance of the cell nucleus to cell division and he observed the structures now known as chromosomes, which contain the genetic identities of cells.

Scientists would later learn how multicellular organisms replace their worn-out cells through cell division called mitosis, named from the Greek word for thread. This is the process whereby two daughter cells are formed with identical chromosomes to the parent cell. Human beings on average experience around 10,000 trillion cell divisions in a lifetime. Unicellular organisms like amoebas use this process for (asexual) reproduction, simply dividing their one cell into two and creating an identical daughter cell: an entire new organism.

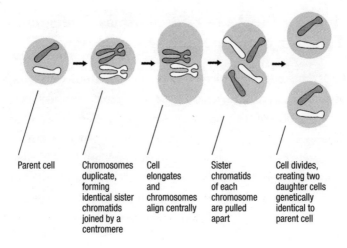

| Parent cell | Chromosomes duplicate, forming identical sister chromatids joined by a centromere | Cell elongates and chromosomes align centrally | Sister chromatids of each chromosome are pulled apart | Cell divides, creating two daughter cells genetically identical to parent cell |

The stages of mitosis.

Sexually reproducing animals and plants use a different method of cell division for procreation, called meiosis, discovered in sea urchin eggs in 1876 by German biologist Oscar Hertwig (1849–1922). In meiosis, offspring are produced from the sexual interaction of the parents: two individual organisms, one male and one female. The chromosomes in the nucleus of a cell from the reproductive organs of each parent duplicate and exchange genetic material, then the cell divides twice, creating four daughter sex cells, or gametes (sperm in the male; eggs in the female), each gamete containing half the number of chromosomes as the original cell. When fertilization occurs, the male and female gametes fuse together, forming a zygote (fertilized egg cell) endowed with genes from both parents. The zygote then divides many times by mitosis, forming new cells and eventually a new offspring.

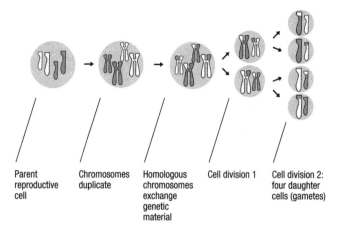

During meiosis reproductive cells divide twice and form four gametes, each with half the number of chromosomes as the parent cell. Human body cells contain twenty-three pairs of chromosomes in each nucleus, while gametes contain twenty-three single chromosomes. After fertilization, and the fusion of two gametes, one of the twenty-three pairs of chromosomes will determine gender. In females the two sex chromosomes are XX; in males, they are XY.

Causing a Revolution: Charles Darwin

Charles Darwin wrote one of the most important and controversial books in scientific history. *On the Origin of Species*, published in 1859, set out his theory of evolution and the law of natural selection. It challenged the belief that God created the universe and all life. Some considered that it challenged the very existence of God, as the process of natural selection appeared to replace the need for a deity or an intelligent creator.

Darwin began the groundwork for his theory during a voyage on the HMS *Beagle* in 1831–6, when he visited the

Galápagos Islands in the Pacific Ocean. He discovered that tortoises on each island exhibited slight physical differences and it occurred to him that they had not in fact been created differently but were developing differences – or evolving – as they responded to the differing environmental conditions of each island. His research proved that the process of evolution had occurred: rather than life on earth being the product of a creator, it had developed from simple to more complex organisms in accordance with their reaction to their surrounding environment.

On his return, Darwin discovered the process that would explain the theory of evolution and revolutionize biology. 'Natural selection' was the cause of evolution, and Darwin described it by analogy with artificial selection. Whereas a breeder can play a modifying role in breeding domestic plants and animals by artificially selecting mates, in natural selection there is no breeder. Instead characteristics such as an aptitude for survival and being well adapted to the environment serve to shape the future of individual species by the natural process of selecting the fittest organisms and eliminating the unfit ones. The idea of competition was employed in Darwin's model to explain the driving force for evolution and phenomena such as extinction and diversification through time.

Alfred Russel Wallace (1823–1913) developed a theory of evolution independently of Darwin, and many scholars credit Wallace as the co-discoverer of evolution. Although their theories were similar, there were differences. For example, where Darwin (correctly) emphasized that selection acted on individuals, Wallace thought it acted on groups or species.

Ideas developed that were loosely related to evolutionary theory but would not have been endorsed by Darwin himself. For example, a year after Darwin's death, his half-cousin Francis Galton (1822–1911) applied the concepts of Darwinism to an idea he called 'eugenics', that of improving the quality of humans born into this world through inheritance of traits considered desirable in human society. Eugenics in the twentieth century became stigmatized after it was taken up in the rhetoric of Nazi Germany in its drive for genetic 'purity'.

Charles Darwin (1809–82)

Born in 1809 into an upper-middle-class English family, Darwin was sent to Christ's College, Cambridge, and then was invited to sail round the world with Captain Robert Fitzroy, who required a naturalist for his scientific expedition.

On board HMS *Beagle*, Darwin studied *The Principles of Geology* by Charles Lyell (1797–1875) and was deeply influenced by Lyell's discussion of James Hutton's view that the earth is much older than biblical scholars had claimed (*see* p. 201). Throughout the many excursions on the trip, Darwin was fascinated by the earth's varied species of animals and plants. He conceived his theory of natural selection in 1838 and devoted the next twenty years to exploring this new idea of evolution. Predictably it drew an angry response from members of the Christian Church.

In 1839 Darwin married his cousin Emma Wedgwood and together they had ten children. As the author of many books on the natural world, he became very famous. He is buried in Westminster Abbey, London.

Investigating Bacteria: Ferdinand Cohn

The bacteria discovered by Antonie van Leeuwenhoek (*see* p. 139) in the 1670s became the curiosity of kings and queens of Europe. But these tiny single-celled microorganisms – no more than a few micrometres in length (a fraction of the diameter of a human hair) – would not be properly understood for another 200 years.

Ferdinand Cohn was one of the first to recognize the existence of different species of bacteria, and in 1872 he published his system of classification dividing them into four groups: sphaerobacteria (round), micro-bacteria (short rods), desmobacteria (longer rods or threads) and spirobacteria (screw-like or spiral). His discovery that different kinds of bacteria have different properties was hugely important in establishing the idea that bacteria can cause infections. With Cohn's support, Robert Koch would go on to find the bacterial causes of anthrax, cholera and tuberculosis (*see* p. 180).

In 1876 Cohn described the entire life cycle of *Bacillus subtilis*. He became the first person to demonstrate that these bacteria form endospores upon exposure to heat. Many bacteria can be killed by boiling, but endospores are resistant to heat; when environmental conditions become favourable again (for example, a return to room temperature), the spores germinate to form new bacilli. Endospores are a problem in today's food industry, where care must be taken to destroy them or use preservation methods that stop endospore-forming bacteria from growing.

Ferdinand Cohn (1828–98)

Born in the German-Jewish ghetto of Breslau, Silesia (now Wrocław, Poland), Ferdinand Cohn was a child prodigy, able to read before the age of two and attending university at age fourteen. However, his first degree was withheld due to his Jewish ancestry and he was unable to obtain a teaching position in Berlin – probably also because of his Jewish roots.

Cohn's interest in the microscopic world developed after his father, a successful merchant, gave him an expensive, high-quality microscope. Cohn was nineteen by this time, with a doctorate in botany and a position at the University of Breslau. His new microscope became one of his treasured research tools. He was appointed associate professor of botany in 1859 and by the 1870s was the foremost bacteriologist of his day, attracting many students and young scientists to his lectures. He is now considered to be a founder of the science of bacteriology.

Father of Genetics: Gregor Mendel

Before the Austrian botanist and monk Gregor Mendel (1822–84) began his work on inheritance, no one understood how tiny units in cells (today called genes) carried traits from one generation to the next.

Aristotle (*see* p. 17) believed that traits were transferred to the next generation through the blood. Likewise the French biologist, the Chevalier de Lamarck (1744–1829), thought it was blood that carried a trait for giraffes to have long necks.

People were also puzzled about which traits showed up in offspring. A popular view was that offspring inherited a blend of the parents' traits; for example, one tall parent and one short parent produced offspring somewhere in between those two heights. But this would imply that over time offspring all converge on an average value for height, which is clearly not the case.

Mendel was curious to unravel this mystery but, born into a farming family, funds for his schooling were scarce. When money ran out, he entered an Augustinian abbey in Brno (now in the Czech Republic), where the abbot supported his university education and encouraged him to carry out plant experiments in the abbey garden. Mendel spent eight years cultivating and cross-breeding tens of thousands of common garden pea plants, tediously counting and categorizing hundreds of thousands of peas. He wanted to see how hereditary traits were transmitted from parent plants to their offspring.

In each generation of plant he compared traits such as stem height (tall or short), colour of flower (purple or white) and seed/pea colour (green or yellow), and he found that offspring plants always showed one or other of each trait, not a blend of the two. For example, the flowers were always purple or white, and not a different colour blend.

Cultivating further generations he found that one of each pair of traits was dominant. For example, in the first offspring the seeds were always yellow (the dominant factor), while in the second the seeds were predominantly yellow in a 3:1 ratio. This ratio appeared in successive generations, too.

His conclusions, published in 1866, are known as Mendel's laws, and although he experimented on pea plants he proposed that these laws applied to all living things. The law of segregation established that there are dominant and recessive traits passed on randomly from parents to offspring, replacing the old idea that inherited traits were a blend of the parents' traits. The law of independent assortment established that traits are passed on independently of other traits from parent to offspring. So, for example, a pea plant with purple flowers is no more or less likely to have yellow rather than green peas.

Explaining Mendel's laws in modern terms, a gene for a particular trait, for example pea colour, can come in different forms, or alleles, which can be either dominant or recessive. Every inheritable trait is determined by a pair of alleles. In the production of gametes (sex cells) during the process of meiosis (*see* p. 147), allele pairs for each trait segregate (separate) so that each gamete carries just one allele for the trait. During fertilization, the gametes carrying single alleles randomly unite, so that offspring inherit an allele from each parent. Whether the trait 'shows' in the offspring depends on whether the alleles are dominant or recessive. Dominant alleles show their traits even when combined with a different kind of allele, whereas recessive alleles only show their traits when paired up with an identical allele. Some inherited disorders, such as cystic fibrosis, are caused by a 'recessive' allele, which means it must be inherited from both parents.

Mendel's results, largely unrecognized during his lifetime, were rediscovered in the early twentieth century and formed the foundation for a revolutionary science: genetics. His theory of inheritance had a profound influence on scientists' understanding of many subjects, including evolution, biochemistry, medicine and agriculture, and it laid the foundations for modern sciences such as genetic engineering.

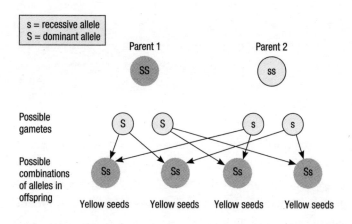

1. Mendel's first-generation cross: the offspring all have yellow seeds, though they carry the recessive allele for green seeds.

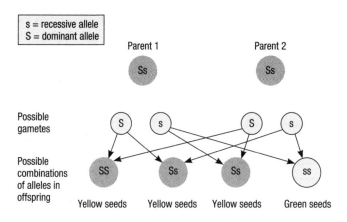

2. Mendel's second-generation cross: most of the offspring have yellow seeds in the ratio 3:1. A quarter have green seeds.

Groundbreaking Genetics: Thomas Hunt Morgan and Barbara McClintock

Following on from Gregor Mendel (*see* p. 152), American biologist Thomas Hunt Morgan (1866–1945) helped to develop the chromosomal theory of heredity in his experiments on the common fruit fly. This is the theory that genes are real, physical objects located within threadlike protein structures known as chromosomes. During meiosis (reproductive cell division) chromosomes exchange segments with one another, a process known as genetic recombination.

The rearrangement of genetic information is an essential process for maintaining genetic diversity. It explains the variation we see between the offspring of one set of parents and it ensures that each generation has new genetic combinations for natural selection to act upon – and to favour those individuals best adapted to the environment. Asexually producing organisms have no such advantage; they rely instead on periodic mutations to provide their variation, making them less responsive to rapid changes in their environment.

Morgan also showed something Mendel had not noticed: some genes located close together on the same chromosome (known as 'linked genes') can be inherited together; in other words, Mendel's second law of independent assortment does not always apply. Linked genes account for conditions like colour blindness being passed down through generations of the same family.

Twenty years after Morgan, the American scientist Barbara McClintock (1902–92) focused her research on the genetic material of maize plants. Comparing the chromosomes of parent and daughter plants, she noticed that parts of chromosomes can switch positions, disproving the accepted theory that genes are fixed in their position on a chromosome. These 'transposable elements', or 'jumping genes', can create mutations or permanent changes to the genetic instructions stored in chromosomes.

McClintock's pioneering research showed how changes in the structure of chromosomes can lead to problems and disease, including cancer, by affecting the instructions stored in cells that govern their normal development and function.

Unlocking the Mysteries of DNA: Francis Crick and James Watson

DNA (deoxyribonucleic acid) was first identified by Swiss biologist Johannes Friedrich Miescher (1844–95) in 1871, but the structure of this essential molecule for life would prove a mystery for another eighty years. DNA molecules are the main constituent of chromosomes, found in the nuclei of plant and animal cells, and their function is to store the genetic instructions, or hereditary information, of living organisms. Genes are simply short sections of DNA.

In the early 1950s English physicist Francis Crick (1916–2004) and American geneticist James Watson (born 1928) joined the race with other researchers to discover the then-mysterious structure of DNA. It was already known that the DNA molecule was made up of four different kinds of simpler units, called nucleotides: adenine, cytosine, guanine and thymine. There was also evidence that the amount of guanine was equal to cytosine and the amount of adenine was equal to thymine.

Using cardboard models of each nucleotide, the two friends tried to puzzle how they fitted together. Watson soon realized that they could only be paired in a certain way: adenine with thymine and cytosine with guanine. They found another crucial clue in the form of X-ray images of DNA made by Rosalind Franklin (1920–58), which were shown to them by their friend Maurice Wilkins (1916–2004) without Rosalind's knowledge. The images suggested a helical structure.

Armed with this data, the pair came up with the correct form of two parallel strands, gently twisted to give the appearance of a double helix. The nucleotide pairs connect the two chains like the rungs of a ladder.

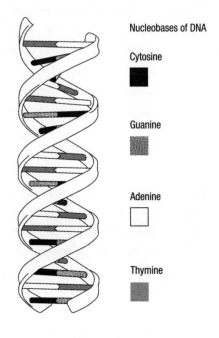

The DNA double helix.

The DNA model they presented in 1953 immediately suggested a possible mechanism for replication and the transmission of hereditary information from one generation to the next: the nucleotides can only be paired in a certain way so the nucleotide sequence on one strand serves as a template for the assembly of a new complementary strand during cell division.

Watson went on to lead the Human Genome Project (HGP) in the 1980s. This international project aimed to decipher the entire human genetic code (genome). In 2000, a 'rough draft' of the genome sequence was jointly announced by the publicly funded HGP and a commercially driven research company set up by American geneticist and businessman Craig Venter (born 1946). The complete sequence was published in 2003. For the first time the human genome, or 3 billion DNA letters representing the code for the genes in a human cell, had been read and put in the correct order. It is now understood that there are around 20–25,000 genes in the human genome (a few more than a chimpanzee).

This result is not of any particular individual's genome, but a combined genome of a group of anonymous donors. All humans have unique gene sequences, and it is because of this that we can have genetic fingerprinting – an invaluable tool used for identification in forensic science since the 1980s.

Hazardous Genetics: Paul Berg

Paul Berg (born 1926), an American biochemist and molecular biologist, invented a way of artificially introducing genetic material from one organism into the genome of another organism. It was the beginning of genetic engineering, which has many applications, some of them controversial.

In the 1970s Berg was studying why cells sometimes spontaneously turn cancerous. He thought that interactions between their genes and their cellular biochemistries were responsible, so he decided he could examine this by introducing a cancer gene into a simple single-celled organism like a bacteria. He felt he could smuggle the gene into the bacteria if he could combine it with genetic material that can normally enter bacteria, such as a bacteriophage, a virus known to infect bacteria. He chose to work with a virus that causes cancer in monkeys (SV40) and the bacterium E. coli, which is found everywhere and is often used in labs.

His 'cut-and-splice' method employed an enzyme to 'cut' the double strands of the DNA from the bacteriophage at exactly the spot that he wanted, and he then used a different enzyme to add sections to just one strand, creating a long 'sticky end' ready to be linked to a similarly treated piece of DNA from the SV40 monkey virus. Then he successfully combined the two molecules, creating a hybrid 'recombinant DNA molecule'.

At this point Berg voluntarily put a stop on the research. E. coli can sometimes exchange genetic material with other types of bacteria, including some that cause human disease. He realized that if he inserted his hybrid DNA into the bacteria, and any escaped and spread, he could not predict the outcome, but it might cause a medical disaster.

In 1974 Berg called for a moratorium on genetic engineering until the dangers could be assessed. The next year, a conference of 100 scientists from around the world agreed guidelines and a ban on any experiments where genetically engineered organisms might be able to survive in humans if they escaped the lab.

Gene therapy and controversial, genetically modified food crops are some of the results of recombinant DNA. Insulin, human growth hormone (the hormone regulating growth in an individual) and some antibiotics are now made using his technique.

Berg is recognized for his work in biochemistry and genetic engineering – and his stance on responsible science.

Catalytic Properties of RNA: Sidney Altman

When Canadian-American molecular biologist Sidney Altman (born 1939) started his research it was not clear how DNA, or genetic information, was conveyed into living cells to instruct the process of growth in an organism. The standard view was that nucleic acids such as RNA (ribonucleic acid) simply carried the genetic codes from the DNA, triggering the creation of enzymes, which in turn catalysed vital chemical and biological reactions in the cells.

Altman and his team discovered conclusively that RNA itself is the catalyst for biochemical development, and therefore RNA can act like an enzyme.

The work of Altman and Thomas R. Cech (born 1947) contributed to our understanding of how life originates and develops. They showed that nucleic acids are some of the basic building blocks of life, acting as both genetic codes and enzymes.

Their discovery of RNA's actions has important and optimistic medical implications. One day enzymes might be applied to cut out infectious or abnormal sequences from the genetic material of a patient suffering from cancer or AIDS.

Campaigning for the Environment: Rachel Carson

American biologist, ecologist and science writer Rachel Carson almost single-handedly started the modern worldwide environmental movement with her powerful book *Silent Spring* (1962), which eloquently showed the devastating effects of pesticide pollution on the natural world. Her 'wake-up call' brought conservation and wildlife organizations together and inspired a new generation of environmental activists.

Carson was the first scientist to point out that pesticides created to kill just one weed or one insect or animal pest have a much wider impact by poisoning the food supply of other species, sometimes killing all insects, birds, fish and wildlife in the area, and lingering on in the soil to have a lasting effect. She called these chemicals 'biocides', and identified more than 200 chemicals developed since the 1940s to kill pests or weeds and widely available for public use in the USA.

Pioneering holistic concepts, she stressed that human beings are also part of nature and our health is harmed by destructive environmental practices just as much as any other species. She showed that pesticides can go on to contaminate the human food chain.

Carson was not automatically opposed to the use of chemicals in agriculture, but she argued that while the long-term effects of newly developed pesticides were not known, it was scientifically and morally wrong to use them indiscriminately and on a large scale. As with the dumping of nuclear waste at sea, she pointed out that there was a lack of research on the long-term effects and 'the mistakes that are made now are made for all time'.

Her work on pesticides sparked an immediate public debate that forced the US government to examine the issue. As a result, a federal government advisory committee in 1963 called for research into the potential health hazards of indiscriminate pesticide use.

Eventually several artificial pesticides, including DDT, were banned in the USA, and many other countries introduced restrictions on their use. She was the inspiration behind the establishment of the Environmental Protection Agency in the USA and she introduced into the public consciousness several concepts, such as 'ecosystem', that are now part of everyday language.

Rachel Carson (1907–64)

Growing up in the small riverside town of Springdale in Pennsylvania, USA, Rachel Carson was encouraged by her mother to appreciate the natural world. Throughout her career as a marine biologist, and later as a general ecologist, her work was just as much her lifetime interest and hobby as a job.

After studying English, then zoology, she spent several years working for the US Bureau of Fisheries, later the Fish and Wildlife Service, being only the second woman to secure a full-time professional job within the service.

She left her job to write *Silent Spring*, the book that woke the world to the dangers of industrial and agricultural pollution. Its title refers to an unnaturally silent region devoid of all nature as a result of artificial chemical pesticides.

Carson worked at a time when Americans believed that science could only be a force for good, and her proof that scientific progress was damaging the environment came as a double shock.

Carson inspired scientific studies of the relationship between environmental pollution and human health. In a horrible irony, breast cancer, the disease from which she died, has since been identified as a cancer that can sometimes be caused by environmental contamination.

CHAPTER 6

The Human Being and Medicine

FACED WITH THE unknown, prehistoric peoples might have turned to spiritual rituals and witch doctors for healing, but they might also have handed down useful healthy practices and knowledge of healing plants. Their burial remains show signs of body preparation and ritual, implying they knew something of bone structure and internal organs. Some remains suggest they even carried out surgery. Evidence of trepanation, or the practice of making a hole in the skull to relieve headaches or mental disorders, has been found in remains from about 10,000 BCE.

Early medical systems existed in China and India, but the oldest documented medical practices are in ancient Egypt and these show significant advances on purely spiritual cures for illness. The ancient Greeks absorbed this knowledge and soon worked out that the body's control centre was the brain, not the heart, as the Egyptians had thought.

After the fall of Rome, as Europe entered the Dark Ages, the Church played a part in stagnating medical progress by prohibiting the dissection of human bodies, though surgeons' skills were tested during this period as a result of the constant fighting. Eventually, ancient scientific and medical knowledge transported via the Arab world would re-emerge in Europe and, with a key discovery that germs cause disease, usher in an era of medical breakthroughs.

First Recorded Medical Practices: Ancient Egypt

Papyrus documents dating to around 1800 BCE show surprisingly advanced medical practices in ancient Egypt. Homer, in the *Odyssey* (c. 700 BCE), remarked that 'the Egyptians were skilled in medicine more than any other art', and Herodotus, a Greek historian who travelled to Egypt around 440 BCE, wrote that their physicians had an excellent reputation and were sought after by other rulers.

The Egyptians believed that gods created and controlled life, and evil gods and demons affected the workings of the body, causing illness by blocking its 'channels', much like a blockage on the River Nile would cause damage to crops. In their view, the bodily channels carried air, water and blood, with the heart in control at the centre. The heart was the seat of intelligence and was treated appropriately: during mummification embalmers left it in the corpse in preparation for the afterlife, whereas the brain was extracted through the nostrils using an iron hook, and rinsed from the skull.

Healers were trained to unblock obstructions in the bodily channels: they examined patients, made diagnoses and offered practical treatments and advice, such as laxatives for a blockage and a balanced diet for general good health. Many recorded treatments would have been ineffective but some, involving types of mould for example, could have provided a cure, though at a risk of infecting the patient.

Egyptian healers could also mend broken bones and stitch and bandage wounds, though they had only herbal antiseptics so surgery would have been risky, not to mention painful!

Medicine as a Rational Science in Ancient Greece: Alcmaeon and Hippocrates

Medical knowledge from Egypt was handed down to the ancient Greeks, who began to develop their own set of beliefs and a system of medicine.

Alcmaeon of Croton (fifth century BCE) was an early pioneer of anatomical dissection. He dissected animals and discovered that the brain controlled sensations, concluding that the brain, not the heart, was the source of sensation and thought. Hippocrates (*c.* 460 to *c.* 370 BCE) became the first recorded person to dissociate the symptoms of illness from religion, magic and superstition, earning him the title of Father of Medicine. He believed that disease was caused by the environment and its symptoms were the natural reactions of the body to the disease. His guidelines for physicians' conduct, professionalism and responsibility in preserving life are the basis of the Hippocratic oath still sworn by medical students today.

Born on the island of Kos, Hippocrates travelled around ancient Greece, practising and teaching medicine, trying to counter the prevailing view that sickness was a punishment from the gods. Hippocrates taught that illness was a result of an imbalance of the bodily fluids known as the four humours: 'Man's body has blood, phlegm, yellow bile and black bile . . . through them he feels illness or enjoys health. When all the humours are properly balanced and mingled, he feels the most perfect health. Illness occurs when one of the humours is in excess, or is reduced in amount, or is entirely missing from the body.'

Hippocrates thought that the body had certain points of 'crisis', which were times during disease progression when the patient would either restore their balance by the healing power of nature, or suffer a relapse.

He instructed physicians to be clean, well kept, honest, kind to their patients, calm, understanding and serious; they should follow specific guidelines on lighting, personnel, instruments and techniques; and of course they should keep clear and accurate records: all guidelines followed today.

Medicine in ancient Greece was divided between the Knidian and Koan (Hippocratic) schools. The Knidian focus on diagnosis was based on many erroneous assumptions about the human body (the Greek taboo forbidding the dissection of humans meant there was little knowledge of human anatomy and physiology). By contrast, the Hippocratic or Koan strategy, focusing on patient care, general diagnoses, knowledge of the likely course of diseases and non-invasive treatments, such as complete rest and immobilization, achieved much greater success.

Dissections and Anatomy in Roman Times: Galen

Galen set the standard for medicine in the Roman world. He lived at Pergamon (now Bergama in Turkey), an ancient Greek city that had become part of the Roman Empire. Such was his following, Pergamon became a centre of medicine in the Mediterranean world during the second century, and Galen's view of the body would prevail in the Middle East and Europe until the seventeenth century.

Galen adopted Hippocrates' theory of the four humours and added his own idea that imbalances could be pinpointed to specific organs or locations in the body. This greatly helped doctors to make diagnoses and prescribe remedies to restore the body's healthy balance.

For much of his life, Galen focused on anatomy, which he understood to be the basis of medical knowledge. Roman law banned the dissection of human corpses so he experimented on pigs, sheep, monkeys and other animals, which occasionally led to mistakes such as his description of the uterus, which is only relevant to dogs.

By severing a pig's spinal cord and tying up the nerve of the larynx he learned how the brain controls the voice, and by tying off a pig's ureter he learned about the functions of the bladder and kidneys. He concluded that there were three connected bodily systems: brain and nerves, heart and arteries, and liver and veins, each responsible for different functions.

Never afraid of trying, Galen carried out operations on the brain and eye, such as cataract removal using a long needle, an ancient technique that could be quite successful if the lens capsule remained intact, or otherwise destroy the eye and cause severe infection. It is remarkable that these operations were done with such inaccurate knowledge of the anatomical position and function of the lens, but Galen would have had more knowledge than others to carry out such a procedure, and his surgical practices followed the strict guidelines of Hippocrates.

In the ninth century Galen's many books were translated into Arabic, influencing the development of Arab medicine, and eventually were translated back into Latin, helping to rekindle a thirst for medical science as Europe emerged from the stagnation of its dark years. His emphasis on bloodletting as a cure-all was continued until the 1800s, and he introduced the still-standard taking of the pulse.

A thousand years after Galen, though restrictions on human dissection remained in place in medieval Europe, Italian Leonardo da Vinci (1452–1519) obtained special permission to dissect human corpses from hospitals in Florence and Rome. His masterly drawings revealed previously unknown anatomical details. Had they been widely published they would have accelerated advances in medieval science and medicine, but Europe wasn't ready for Leonardo. Another 200 years would pass before knowledge of human anatomy and bodily functions could make a real impact on medical science.

Galen (129 to *c.* 216 CE)

Galen was the son of a well-off architect. He received a good education in medicine in Pergamon, where there was a famous temple to the god of healing, Asclepius. Later he worked as a doctor in a gladiator school and learned a great deal about open wounds and physical trauma.

Ambitious and clever, Galen moved to Rome in 162 CE, working his way up to becoming a physician to the emperors Marcus Aurelius, Commodus and Septimius Severus.

Golden Age for Medicine in the Arab World: Rhazes

As Europe declined after the fall of Rome, Islamic countries flourished during a golden age of scientific, cultural and economic development. In Iran, a Muslim doctor known as Rhazes (854 to *c.* 925 CE) became the Islamic equivalent of Hippocrates (*see* p. 167) – though he wasn't afraid to challenge the authority of the great ancient Greek physician. Rhazes helped to show that diseases have organic causes and are not due to magic, fate or supernatural powers; he also wrote the first book on childhood illnesses, earning him the title of Father of Paediatrics.

Rhazes started out as either a jeweller or a money-changer; he was also a musician and an alchemist. He only became interested in medicine when an alchemical experiment blew up in his face and damaged his eyesight, and at the age of thirty he began medical and philosophical studies in Baghdad, then a centre of Islamic science. He soon became a famous doctor, writing more than 100 medical texts, and his continuing interest in alchemy, in those days considered to be just another natural science, undoubtedly contributed to his skill as a doctor, since it was from that practice that he first learned the empiricism he brought to medicine.

He was the first known doctor to discover that smallpox and measles are different diseases; he had the insight that some fevers are the body's defence mechanism to fight infection; he recorded the first known use of animal gut for sutures; and he was the first to use plaster of Paris for casts. He was also one of the first practising doctors to discuss medical ethics and the reasons why people choose to put their trust in a particular doctor.

Rhazes lived at a time when treatments for disease were sometimes difficult to explain. According to one story, he was called to treat an emir who was so crippled by arthritis that he could not walk. Rhazes ordered the man's best horse to be brought to the door, before treating the patient with hot showers and a potion. Then he pulled out a knife, swore at the man, and threatened to kill him. The emir jumped to his feet and charged at the doctor, who fled for his life to the waiting horse. When he was sure he was safe he wrote to the emir, explaining that his treatment had softened the humours (bodily fluids) and he had left it to the patient's own temper to finish dissolving them.

Rhazes gave so much to charity that he died poor himself. According to legend he suffered from an eye cataract towards the end of his life but refused to have any treatment, saying that he had seen so much of the world he was tired of it.

Middle Eastern Medical Textbook: Avicenna

Avicenna, a brilliant medieval philosopher–scientist and doctor from Bukhara (now in Uzbekistan), included much valuable medical knowledge of the ancients in his *Canon of Medicine*, along with knowledge from Mesopotamia and India, and his own discoveries. His fourteen-volume encyclopaedia became a standard medical textbook in the Islamic and Christian worlds.

Throughout his career, Avicenna stressed the need for empirical medicine: examining, testing and not taking anyone's theory for granted without proof. He describes his approach here: 'In medicine we ought to know the causes of sickness and health. And because health and sickness and their causes are sometimes manifest, and sometimes hidden and not to be comprehended except by the study of symptoms, we must also study the symptoms of health and disease.'

Among his many investigations were: the contagious nature of some diseases; the impact of environment and diet on health; the spread of diseases in water or soil; and diseases of the nervous system causing mental disorders. He supported clinical trials of medicines and his rules covered much the same ground as they do today, such as using a large enough test group to ensure results are not just accidental. He also believed that experiments should be carried out on humans, not animals, because 'testing a drug on a lion or a horse might not prove anything about its effect on man'.

He had ideas on quarantine to contain the spread of infections and he speculated on the existence of microorganisms more than 600 years before Antonie van Leeuwenhoek discovered bacteria using a microscope (*see* p. 139). He described sexual diseases, skin diseases, the anatomy of the human eye, facial paralysis and diabetes. Perhaps because of his philosophical training, he was interested in psychology and exploring the effect of the mind on the body. And he is known to have carried out at least one operation: on a friend's gall bladder.

Though famous and much in demand, Avicenna showed great compassion and would treat the poor without expecting any payment.

Avicenna (*c.* 980 to 1037)

Avicenna had memorized the Qur'an and other classic Islamic texts by the time he was ten. He soon surpassed his teachers, began his own course of study in medicine, and was still young when he successfully treated the Sultan of Bukhara.

It was a time of turmoil and Avicenna's life was profoundly affected by political uncertainties. Not only were Turkish tribes displacing Iranian rulers in central Asia, but local leaders in Iran were throwing off the central control of the Abbasid caliphate based in Baghdad.

In 999 the Bukharan ruling family was overthrown by Turkish invaders and Avicenna embarked on a long period of wandering around Iran, which was not without adventure. He escaped a kidnap attempt, hid from arrest and imprisonment, and made a breakneck flight in disguise. In spite of this he wrote philosophical papers, and whenever he settled in one place for long enough he practised his profession.

In about 1024 Avicenna finally found refuge as doctor and advisor to the ruler of Isfahan, remaining in his service for the remainder of his life.

Herbalists: Ibn al-Baitar, Garcia de Orta and the Monasteries

Medicines in the Middle Ages, like those in ancient times, generally came from plant sources. Ibn al-Baitar (*c.* 1197–1248) was an important herbalist (botanist) during the Islamic golden age. His extensive encyclopaedia of the medical uses and properties of plants was unmatched in Europe and the Middle East for centuries.

Al-Baitar was born near Malaga in Spain, but the Christian reconquest disrupted the area and, like thousands of other Muslims, al-Baitar emigrated. Soon after 1224 he settled in Egypt, where he became chief herbalist to the ruler al-Kamil and in this role he collected plants from Palestine, Arabia, Greece, Turkey, Armenia and Syria.

He had an incredible memory for plants and their ancient medical uses, and he obtained new remedies 'through experimentation and observation', always extensively testing his drugs. His *Book of Simple (Herbal) Remedies* was a systematic compilation of the medicinal and general properties of 1,400 different plants, including 200 previously unrecorded. In a later work he focused on medicinal cures, listing plants according to the diseases of the body that they could help treat: ear, head, eye and so on.

Garcia de Orta (*c.* 1501–68), a Portuguese Renaissance physician of Jewish descent, used an experimental approach to plant medicines while working in the Portuguese colony of Goa, India. His book on herbs and drugs introduced Indian medicinal plants and Eastern spices to Europe; he also passed on knowledge of tropical diseases, including the Asiatic form of cholera (the infectious disease of the small intestine).

Elsewhere in medieval Europe, monasteries served as repositories of learning and monks busied themselves translating and copying ancient works from the Arab, Greek and Roman worlds. They discovered herbal remedies for common ailments among the ancient writings and developed herb gardens so they could supply the raw material for their growing medical centres. Village healers also prescribed herbal remedies, sometimes adding their own spells and enchantments, which often led to accusations of witchcraft.

Modern scientists have shown how some ancient herbal remedies were successful, for example willow bark used to relieve headaches 2,000 years ago contains salicylic acid, the active ingredient in aspirin. Other treatments proved less successful: a medieval cure for baldness involved rubbing an onion on the affected scalp. Herbal remedies also proved ineffectual in the face of epidemics like the 'Black Death', the deadly plague that swept across Europe between 1346 and 1353, spread by fleas on rats aboard merchant ships.

In the nineteenth century, chemists began to extract the active ingredients from plants, and so dawned the era of chemical drugs. These have now displaced plants and herbs in standard Western treatments, although compounds derived from plants are still used in modern medicine.

Pioneering Vaccines: Edward Jenner

Although Antonie van Leeuwenhoek (*see* p. 139) first observed microorganisms in the 1670s, it took many years for people to understand that microorganisms (or 'germs') cause disease. The desire to control the causes of infectious diseases was a great incentive for learning more about these living organisms.

Edward Jenner (1749–1823), an English surgeon and nature enthusiast, was one of the pioneers who helped people appreciate the link between germs and disease, and his research on inoculation launched a new chapter in medicine: immunology.

Smallpox was one of the most feared diseases of Jenner's time. It particularly affected infants and small children, with a high death toll and the horrible disfigurement of any survivors. Blessed with an inquisitive mind and a natural instinct, Jenner was convinced that the cowpox and swinepox viruses affecting animals could be linked to the human form, smallpox, and he hoped the link would lead to a cure.

He had heard folk stories of milkmaids who had suffered from cowpox and then appeared to be immune to smallpox. In 1796 he applied this hypothesis to eight-year-old James Phipps by inoculating him with pus from the wounds of dairymaid Sarah Nelmes, who had caught cowpox from a cow named Blossom. Jenner used a stick to transfer the diseased pus from the wound in Sarah's arm and place it directly into a suture on James. Apart from initial symptoms of fever and general discomfort, James Phipps did not contract smallpox. Jenner tested the success of the vaccine further by inoculating James with smallpox material, and found similar results, proving that immunization was successful.

The Royal Society, ever cautious, did not publish Jenner's work for several years, saying, at first, that there was insufficient evidence to support such a revolutionary finding. Despite initial public disapproval, Jenner carried on vaccinating patients, including his own eighteen-month-old son. Eventually the reality of his work and the results of his vaccine won over his critics.

Although the farmer Benjamin Jesty (*c.* 1736–1816) successfully inoculated his family with cowpox in 1774, some twenty years before Edward Jenner, it is thought that Jenner arrived at the same findings independently and added value to them with his experiments and explanations.

As a consequence of Jenner's groundbreaking work, smallpox was declared an eradicated disease by the World Health Organization in 1980. It is no easy task to eradicate a disease from the world; to date, smallpox is the only infectious disease affecting humans that has been eradicated, helped by the fact that it is easily recognized from the outset (a rash). But it does raise hope that the same can be accomplished for other diseases, such as polio, which although eliminated in many countries, still persists in some areas.

Disease of the Cells: Rudolf Virchow

The German physician Rudolf Virchow (1821–1902) promoted the theory that disease originates in cells or is represented by cells in an abnormal state. 'Think microscopically' was his persistent admonishment to his medical students in his encouragement of their use of microscopes. His work provided the foundations for modern pathology – the science of the causes and effects of disease.

Virchow's approach to disease at the cellular level led him to pioneering research in oncology. Not only was he the first to correctly describe a case of leukaemia (cancer of the blood), but he found other types of malignancy, too, including gastric cancer (cancer of the stomach); one of its symptoms, characterized by an enlarged supraclavicular lymph node, is now known as 'Virchow's node'.

In 1848 Virchow conducted a study into the typhoid epidemic in Silesia, and argued that the country's poor health standards and disease-ridden populace were a direct result of their lack of freedom and democracy. It was the starting point of his theories linking practical medicine and political legislation, which led him to declare: 'Medicine is a social science, and politics is nothing but medicine on a large scale. The physicians are the natural attorneys of the poor, and the social problems should largely be solved by them.'

His comprehensive vision, and recognition that disease so often results from poverty, earned him the title 'Father of Public Health'.

The Dangers of Microbes: Robert Koch and Louis Pasteur

The son of a German mining engineer and one of thirteen children, Robert Koch (1843–1910) taught himself to read with the help of a newspaper. He went on to contribute pioneering research that would help to identify and isolate microbes responsible for disease.

It was a time of major anthrax pandemics across Europe, and this disease carried by goats, sheep and cattle was an occupational hazard for animal workers. Although the anthrax bacillus (disease-causing bacterium) had been described by the French physician Casimir Davaine (1812–82), no breakthroughs for its prevention and treatment had been found and researchers were at a loss to explain why cattle contracted the disease not only from infected cattle, but also from grazing the pastures where infected animals had been kept years earlier.

In 1875 Koch managed to isolate and culture the bacilli responsible. He observed their entire life cycle, noticing that they formed resistant endospores in unfavourable conditions, such as when there was insufficient oxygen. The spores would lie dormant until the right conditions prevailed, then gave rise to new bacilli, explaining the recurrence of the disease in pastures that had lain fallow.

The French chemist Louis Pasteur (1822–95), famous for discovering pasteurization, also tackled the problem of anthrax. He advised farmers to keep their animals away from contaminated land where the afflicted had died, and in 1877 started work on a vaccine containing anthrax bacteria. Heating his prepared vaccine to 42°C (108°F) to weaken the bacteria, he then injected it into sheep, which contracted, and soon recovered from, a mild case of anthrax, giving them immunity to future attacks.

On 5 May 1882 Pasteur conducted an experiment that garnered quite a crowd. He inoculated twenty-five sheep and left twenty-five without inoculation. Twenty-six days later he injected all fifty sheep with full-strength anthrax bacteria. Two days after that, all the sheep that hadn't been inoculated were dead and all the inoculated ones were alive.

Koch went on to identify microbes that cause tuberculosis and cholera, and Pasteur found a vaccination for rabies. Pasteur tested his rabies vaccine on an afflicted boy, Joseph Meister, in 1885. Ten days later the boy was well again.

Although it was Louis Pasteur who introduced the idea that a microorganism might be cultured outside the body, it was Koch who perfected the technique of pure culture. This begins with a sample containing many species of microorganism, from which similar cells are transferred onto a new, sterile growth medium, and the process is repeated until, through diluting and separating the cells in successive samples, a sample containing just one species of microorganism is obtained.

In the nineteenth century, many people died after operations. Both Koch and Pasteur recognized that a lack of cleanliness was a contributing factor because microbes could invade the body, causing disease and infection. In 1878 Pasteur announced before the Academy of Medicine in France:

> If I had the honour of being a surgeon, impressed as I am with the dangers to which the patient is exposed by the microbes present over the surface of all objects, particularly in hospitals, not only would I use none but perfectly clean instruments, but after having cleansed my hands with the greatest care, and subjected them to a rapid flaming, which would expose them to no more inconvenience than that felt by a smoker who passes a glowing coal from one hand to the other, I would use only lint, bandages and sponges previously exposed to a temperature of 130–150°C [266–302°F].

Pasteur's statement has become the basis of aseptic surgery, which aims to prevent access of harmful germs to the operating space rather than trying to eliminate them with antiseptics applied to the tissues. The advice for surgeons to flame their hands before operating reflected a routine procedure in

Pasteur's laboratory until 1886; he was also known to clean glasses, plates and silverware with his napkin before every meal, such was his concern about germs.

Treating the Brain in the Modern Era

Treating brain injuries and disorders, though practised from the late Stone Age, has proved one of the most challenging areas of medicine in the modern era.

Understanding the brain and how it works was complicated by French philosopher René Descartes (*see* p. 63) when he claimed that nerves contained 'animal spirits' and the mind and body were separate entities. In a deal with the Pope to obtain human bodies for dissection, Descartes is said to have declared: 'Anything to do with the soul, mind or emotions, I leave to the clergy. I will only claim the realm of the body.' It led to Cartesian dualism, the idea of a non-physical mind (or soul) separate from the material body.

Descartes was proved wrong about 'animal spirits' in nerves when English scientist Richard Caton (1842–1926), experimenting on dogs and monkeys in 1875, discovered varying electrical currents in the brain; today the communication of neurones (nerve cells) via electrical and chemical signals is well established. Scientists also failed to find a scientifically proven meeting point between a non-physical mind and the physical body, where the two might interact, so they concluded that the mind is not separate from the body and favoured a physical description of consciousness: the approach of modern neuroscience.

Brain injuries revealed further workings of the brain. In 1848 American railway worker Phineas Gage survived an iron rod driven through his head, destroying much of the left frontal lobe of his brain. It led to an understanding that parts of personality are controlled in the frontal lobe and, indirectly, to the controversial procedure known as lobotomy, used to treat depression and mental illness. This involves cutting connections to and from the frontal lobe, a popular procedure in the early twentieth century when the number of patients residing in mental hospitals ('lunatic asylums') had led to serious overcrowding. By the 1950s, the procedure had largely been replaced by antipsychotic drugs.

Experiments on animal and human behaviour by the Russian physiologist, surgeon and psychologist Ivan Petrovich Pavlov (1849–1936) revealed how the brain reacts to stimuli. He famously noted that salivation is a reflex action in a dog when it is offered meat (just as a hungry person's mouth might water at the sight of a tasty dish). He rang a metronome each time he gave food to the dog, and he found that when he removed the food and simply rang the metronome the dog still salivated – it had learned to respond to a 'conditioned stimulus'. Pavlov concluded that conditioned reflexes are caused by physiological events – the formation of new reflexive pathways in the brain's cortex.

Despite such discoveries, the path of the nervous system remained unknown until the work of Spanish physician Santiago Ramón y Cajal (1852–1934), who became one of the founding fathers of neuroscience. Building on staining techniques pioneered by Italian Camillo Golgi (1843–1926),

Cajal applied a silver stain to sections of brain tissue so he could inspect a single neurone (nerve cell). He found that a neurone has a cell body with outgrowths (dendrites and axons), along which impulses are conducted from one cell to another. The nervous system is not individual neurones arranged in a continuous single network, as Golgi had thought, but individual neurones that are interconnected with other (target) cells through synapses, or structures that allow an electrical or chemical signal to pass from one cell to another.

Cajal's findings helped to redefine our understanding of brain circuits and laid the foundation for subsequent studies of tumours of the brain and spinal cord.

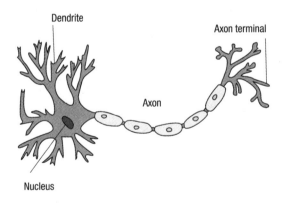

A typical neurone, or brain cell. Axons send electrical impulses from one neurone to another through synapses; dendrites receive impulses from other neurones, also via synapses. There are more synapses in a human brain than stars in the Milky Way.

In the 1970s American neuropharmacologist Candace Pert (1946–2013) shot to fame when she discovered a biochemical basis for consciousness. She had found the brain's opiate receptor – the brain site where endorphins (protein-like molecules acting as the body's painkiller, or 'bliss-makers') bond with brain cells. In other words, the chemicals in our bodies are also involved in emotion, showing the brain and the body to be fully integrated as one system at the molecular level: neither can be treated separately without the other being affected. The connection between mood and health, understood by the ancient Greeks and many indigenous cultures worldwide, had finally been re-established in twentieth-century Western medicine.

First Ever Antibiotic: Alexander Fleming

In the 'pre-antibiotic age' even the tiniest scratch could prove fatal, with more soldiers dying from infections than from the wounds themselves. This all changed in 1928 when Alexander Fleming, a Scottish biologist and bacteriologist, identified the bacteria-killing properties of penicillin, the first ever antibiotic and one of the greatest discoveries in medicine. It was the beginning of a golden age of medical discovery that would save the lives of millions of human beings.

Fleming's job at St Mary's Hospital, London, was to search for new ways to protect against germs; it involved growing bacteria, then killing or weakening them with chemicals, and testing the resulting vaccines. His great discovery came about by accident when he left a Petri dish of live bacteria uncovered in his lab while he went on holiday. On his return, there were many patches of different moulds growing on the dish, and he was observant enough to spot that one in particular seemed to have killed the bacteria around it.

Moulds grow from microscopic, seed-like spores, which float in the air and are difficult to guard against. Researchers normally throw away mould-contaminated specimens, but fortunately Fleming kept the sample and soon identified the white, fluffy growth as a type of penicillium, common on soil, rotting fruit and rotting bread. He extracted some 'mould-juice' (penicillin) and many tests later found that penicillin killed or stopped the growth of even the most harmful bacteria.

He adopted the term 'antibiotic', meaning 'against life', for the penicillin that he tried to purify from mould-juice, but it was not until 1939, after the invention of vacuum freeze-drying, that the Oxford biochemists Howard Florey (1898–1968) and Ernst Chain (1906–79) isolated pure penicillin.

Fleming's discovery completely changed the lethal risk of everyday accidents and probably no other group of drugs has saved as many lives as antibiotics. There are now more than 8,000 different antibiotics available to fight infections and diseases caused by bacteria, such as chest infections, meningitis and tuberculosis. But even in 1946 Fleming noted that some strains of bacteria change so quickly that they can become resistant to antibiotics, particularly if the dose is too small or is stopped too soon, thus predicting today's 'super bugs'. Antibiotics do not harm viruses so they have no effect on infections like the common cold.

Fleming never received patent payments for his discovery of penicillin. When American drug companies collected $100,000 for him in recognition of his contribution to science, he passed it on to his medical school for research.

Alexander Fleming (1881–1955)

Fleming was the son of a Scottish farmer, and his first job was as a clerk in a shipping office in London. But in 1901 he inherited the then large sum of £250 from an uncle and decided to start a new career as a doctor.

After graduation he worked as a bacteriologist in the inoculation department of St Mary's Hospital, London, and also became a surgeon and a medical writer. His quiet sense of fun led him to grow some bacteria cultures just for their pretty colours and patterns. He called them his 'germ paintings'.

During the First World War he and his colleagues went to France as part of the Army Medical Corps, working in battlefield hospitals, but with peace restored in 1918 he returned to the lab. He became professor of bacteriology, and just a few weeks later went off for a family summer holiday. It was when he returned to his lab that he made his legendary discovery of bacteria-killing mould.

A shy, retiring man, his presentation was poor, and his discovery was ignored by other scientists for years, but he would eventually receive full credit.

> Publicized as a 'miracle drug', pure penicillin was mass-produced in 1943, and it immediately saved the lives of thousands of soldiers in the Second World War.

Molecular Diseases: Linus Pauling

Fascinated with the structure of molecules, the American chemist Linus Pauling (*see* p. 130) focused his research after the Second World War on the structure of large biomolecules. These organic molecules (hydrocarbon compounds) form essential parts of living organisms. His investigations led to the discovery of the first 'molecular disease': sickle-cell anaemia.

Pauling had learned from a medical expert that sickle-cell anaemia was caused by the fact that the sufferers' red blood cells were twisted from normal disc-shapes to sickle-shapes, so Pauling set about examining the contents of red blood cells – haemoglobin. After a year, he and his team made an astonishing discovery when they applied an electric field to separate haemoglobin molecules by their electric charge. Sickle-cell haemoglobin molecules had more electric charge than normal haemoglobin molecules. That a potentially deadly illness could be caused by such a slight difference in the molecules gained great attention, and led the way for important research into 'molecular diseases'. Work by Pauling's colleagues later showed that the disease was inherited, cementing a vital link between the fields of molecular medicine and genetics.

Pauling believed that molecular diseases could be treated once an understanding of the molecular structure of the disease was reached. This emerging field of medicine combines modern medical practices with biochemistry and aims to treat disease at a molecular level.

He also gave the term 'orthomolecular medicine' to the concept that optimal physical and mental health can be acquired by making sure 'the right molecules in the right amounts' are in the body. He believed that if the body has the right balance of chemicals, the necessary chemical reactions for health can be optimized.

In a test on himself, he found he had fewer colds when he took large doses of vitamin C. Since his bestselling book on the subject (which attracted criticism from other scientists), studies on the effects of dietary supplements have been ongoing and dietary supplements are now a major industry.

First Vaccine Against Polio: Jonas Salk

American parents in the 1940s and 1950s were terrified by the growing incidence of poliomyelitis or polio. This virus attacked the nervous system and crippled or killed about one child in 5,000 in annual summer epidemics in America.

Jonas Salk (1914–95), an American microbiologist of Orthodox Jewish-Polish descent, discussed the problem in a couple of articles, and the prestigious National Foundation for Infantile Paralysis (now the March of Dimes Foundation), attracted by his obvious energy and enthusiasm, offered him nearly all its research funds to find a cure.

This was the beginning of the scientific community's dislike of Salk. Researchers like Albert Sabin (1906–93) had spent many years in careful research, and suddenly a newcomer to the field had been given seemingly limitless money.

At the time, most vaccine-hunters were working with 'live' but weakened polio viruses, thinking that mild exposure to the disease was the only way to bring about immunity. From his work on a flu vaccine, Salk knew that a 'killed' or deactivated virus could sometimes act as an antigen, triggering the body's immune system to generate antibodies that would attack and destroy any future invasion by that virus, while avoiding the obvious risk of infecting the patient. His most important insight was to apply the same principle to polio and try to find a vaccine based on a 'killed' virus.

Salk used formaldehyde to kill the virus but leave it sufficiently intact to spark the immune system into action. He first tested his vaccine on monkeys, then on a small group of humans. The results showed the successful production of antibodies with no unwanted ill effects. The next stage was a large-scale test programme on children, launched in 1954, which by April 1955 showed that his vaccine was effective and safe.

Normally, scientific results are first published in academic journals before being announced to the world at large, but in 1955 Salk agreed to a press conference before publishing, and although he did not claim personal credit, he found himself a media and public darling overnight. Unfortunately it also meant that he was a villain to many other scientists who felt that he had not given enough credit to other researchers in the field.

In the minds of ordinary people, Salk will forever be the man who beat polio. He particularly endeared himself to the public by refusing to patent the vaccine or profit from it personally.

In 1958 Sabin's vaccine based on a 'live' virus was introduced. This could be taken orally, unlike the Salk vaccine which had to be injected, and it needed fewer booster shots, so it began to replace Salk's 'killed' virus vaccine. Today, the two are usually used in tandem.

The Salk Institute for Biological Studies is now a famous and prestigious institution for molecular biology and genetics.

Therapy for the Mind: Sigmund Freud

Born in Freiberg in Austria-Hungary (now Příbor in the Czech Republic), Sigmund Freud (1856–1939) was the oldest child of seven and favoured for his intellectual brilliance. With a remote, authoritarian figure for a father, and a caring and nurturing mother, his family circumstances played a major part in the theories of the mind that he later formulated.

Unable to make a decent living as wool merchants, the family moved to Vienna. Freud studied medicine and specialized in neurology, studying under Jean-Martin Charcot (1825–93), who used hypnotism to treat hysterical disorders. Freud soon realized that the standard treatments of electrotherapy or hypnotism were ineffective, and instead experimented with the 'talking cure', encouraging patients to talk about and release their problems by confronting them rationally, then making necessary changes to their behaviour. He approached their neuroses through free association of ideas and the interpretation of dreams, which he thought offered an insight into the unconscious. While working he used cocaine to expand his mind, and at some points in his life he was clearly addicted to the drug.

In 1900 Freud published his first approach to psychoanalysis. His central theory was that the unconscious mind drives human behaviour, and psychological distress is created when social conventions override primitive impulses, causing tension and repression. One way to remove the tension and uncover the unconscious wishes or motives is through dreams, which offer a route to self-knowledge.

In his forties Freud intensively explored his own psychology and reached several universal conclusions, particularly that the sexual impulse is the source of many neuroses. The wider scientific community reviled his explorations of sexuality, especially his belief that even infants are driven by it, and for a time Freud had to work in isolation. But by 1906 he had gathered a group of followers, including Carl Jung (1875–1961) and Alfred Adler (1870–1937), and in 1908 the first psychoanalytic conference took place in Salzburg, shortly followed by the establishment of the International Association of Psychoanalysts in 1910.

When Hitler's Nazis took power in Germany in 1933, Freud's books were among the first to be thrown on the public bonfires. Five years later the Nazis took over Austria and began to harass everyone with Jewish ancestry, including Freud, even though he was an atheist. Freud decided he would prefer to 'die in freedom', so he and his family left for London in 1938.

Freud endured many unsuccessful operations for throat cancer, but eventually he persuaded his friend, the doctor Max Schur, to help him die – Schur gave him three doses of morphine, and Freud died peacefully in his home in north London.

Reproductive Medicine: Gregory Goodwin Pincus

All kinds of methods of birth control have been used through history, from honey and acacia leaf pessaries in ancient Egypt, to plants with contraceptive properties in ancient Greece, to elephant dung pessaries in Persia in the tenth century. However, the Catholic Church forbade efforts to prevent pregnancy in medieval Europe, and many 'witches' were punished for carrying out abortions or supplying herbal contraceptives.

In the nineteenth century, the management of population growth became a political issue as controlled birthrates were linked to higher standards of living and greater economic stability.

The invention of the contraceptive pill in 1951 by American Gregory Goodwin Pincus (1903–67) marked a revolution in family planning, while helping to curb the problem of overpopulation throughout the world. It had an impact on women's health, feminist issues, fertility trends, religion and politics, and of course relations and sexual practices of adults and adolescents – a scientific breakthrough in every sense, reaching into virtually all aspects of our social lives.

Chapter 7
Geology and Meteorology

THE HISTORY OF the origin of the earth has puzzled the greatest minds for millennia. Only two or three centuries ago, most people thought that our planet was a mere 6,000 years old.

Progress was made in the eighteenth century when a Scottish scientist and gentleman-farmer, James Hutton, showed how an examination of the layered rocks on the planet could reveal what had happened in the past, and ultimately, the age and origin of the earth. He found evidence that the earth was very much older than suggested in the Bible. Today we estimate that the earth solidified and became a planet some 4.6 billion years ago.

Mining of the earth's crust for its valuable metals and oil shale has gone on since the beginning of civilization, with important advances in Roman times and during the Industrial Revolution. Accurate knowledge of ores and their natural distribution became vital for commercial mining activities, making geology a popular subject of study. Today geologists can systematically and accurately identify ore deposits in the earth's strata around the world.

Earthquakes and floods are just some of the events in the history of the earth for which geologists can find evidence, and knowledge of past earth events is now used to predict events in the future, both here and on other planets.

Like geology, meteorology, or the scientific study of the earth's atmosphere, has ancient origins: civilizations throughout history have needed to forecast the weather. In recent years understanding and predicting changes in the earth's atmosphere, and the impact of climate change on our planet and society, have become matters of extraordinary international importance.

The Ancients' View of Earth: Aristotle and Theophrastus

In the fourth century BCE the ancient Greek scholar Aristotle (*see* p. 17) was thinking geologically when he wrote about the slow changing of the earth over long periods of time: 'If the sea is always advancing in one place and receding in another it is clear that the same parts of the whole earth are not always either sea or land, but that all this changes in course of time.'

In his work *Meteorology* he made some of the first recorded observations on the water cycle. Water vapour, he said, 'rises from hollow and watery places, so that the heat that is raising it, bearing as it were too heavy a burden, cannot lift it to a great height but soon lets it fall again'. Aristotle is now considered to be the founder of the science of meteorology.

Theophrastus (*c.* 371 to *c.* 287 BCE), Aristotle's successor at the Lyceum school in Athens, made one of the first classifications of different stones according to properties such as hardness and behaviour when heated. His treatise, aptly named *On Stones*, also recorded some practical uses of minerals in the ancient world, for example, to make glass, paint or plaster.

Land Formations in China: Shen Kuo

In medieval China, Shen Kuo (1031–95), famous for his concept of true north (*see* p. 25), wrote the first known hypothesis of the way that land structures are formed (later called geomorphology) and the study of fossil plants to indicate climate changes (palaeoclimatology).

On a visit to the Taihang and Yandang mountains he noted that although these land forms were hundreds of miles from the ocean, fossil shells could be found there in a particular geological stratum, or level. He concluded that at one time the area must have been either seashore or under water, the sea having since shifted, leading him to hypothesize that the continent must have been formed over an immense timespan for all the sediments to build up. This predated James Hutton's groundbreaking work on sedimentary deposits by some 650 years (*see* p. 201).

Shen made a further discovery in about 1080 when he found that a landslip on the bank of a river near modern Yan'an (northern Shaanxi, China) had revealed a large underground cavern containing hundreds of fossilized bamboo plants, even though bamboo did not grow in the dry habitat of that region. Ahead of his time he commented on the possibility of climate change: 'Perhaps in very ancient times the climate was different so that the place was low, damp, gloomy and suitable for bamboos.'

Mining and Minerals: Agricola

In sixteenth-century eastern Germany, a brilliant chemist called Georgius Agricola wrote a textbook on metals and mining, *De Re Metallica*, the finest book on metallurgy and mining techniques in the Western world since Pliny the Elder's *Historia Naturalis* of c. 77 CE.

Expensive to produce and with distribution limited, some copies of the weighty tome were chained in churches where the priest would translate the Latin text on request. By 1700, more than a dozen editions had been published in German, Italian and Latin, with a first English translation in 1912 made by a mining engineer and future president of the United States, Herbert Hoover, and his wife, Lou Hoover, a geologist and Latin scholar.

Agricola's definitive work is believed to have taken around twenty years to complete. The collection of twelve books covered detailed aspects of the processes and problems involved in mining operations, and other related areas, including: administration, assaying, construction, miners' diseases, geology, marketing, prospecting, refining, smelting, surveying, use of timbers, ventilation and water pumping – water removal from mines was a major problem at the time.

Agricola also wrote the first scientific classification of minerals by geometric form in *De Natura Fossilium*, an invaluable guide for any budding mining engineer.

Georgius Agricola (1494–1555)

Georgius Agricola is the Latinized name of Georg Bauer – 'bauer' means peasant or farmer. Born at a time when the Renaissance was well established in Europe and the invention of the printing press had brought about an expansion in literacy and a thirst for knowledge, the new ways of thinking were clearly an inspiration to Agricola.

He excelled at school and studied medicine at the University of Leipzig, where he was awarded a degree in 1517 – the year Martin Luther began the Protestant Reformation against the Roman Catholic Church in Wittenberg.

Agricola was appointed physician at the town of Joachimsthal, in the centre of one of the major mining and smelting works in Europe, which enabled him to observe mining techniques and the treatment of ores while he practised medicine. After three years he left the town to tour and study mines around Germany, settling in Chemnitz in Saxony, a prominent mining centre. By the time he had published several works on mining and mineralogy, most of Germany had converted to the Protestant cause. Agricola remained a faithful Catholic all his life and had to resign his office of town burgher because of protests. It is thought that he died while having a heated argument with a Protestant.

Religious and Scientific Struggles: Nicolaus Steno

Christian beliefs were a heavy influence on Western ideas about the origin of the earth in the seventeenth century. English priest and mathematician William Whiston (1667–1752) spent a good deal of time trying to find scientific explanations for biblical stories. He proposed that a comet had hit the earth and caused Noah's Great Flood, whose waters in turn had shaped the earth's geography. He also predicted that the earth would end in 1736 after a collision by another comet, but it didn't happen, much to the relief of the general public.

Caught in the struggle between religion and science was a Danish pioneer, Nicolaus Steno (1638–86). Having made a name for himself as an anatomist, he was sent the head of a huge shark, caught near a coastal town in northern Italy, for dissection and analysis. Steno noticed that its teeth resembled stony objects embedded in rock layers (strata), which led him to propose that fossils were the remains of living organisms from many years ago, preserved in layers of rock. Others, including Robert Hooke (*see* p. 141), had come to the same conclusion, but Steno went further. He proposed that many strata are the result of sedimentation (a build-up of deposits of particles that eventually consolidate to form rock) and that a study of the fossils embedded in different strata could reveal a chronological history of the earth's geological events. This was revolutionary, as was his idea that mountains are formed by changes in the earth's crust – not simply structures that grow from the earth like trees, as previously thought.

Steno had made a great step forward but he grossly underestimated the timespan for the whole of the earth's geological history, choosing to adopt the accepted view of 6,000 years, as derived from the Bible.

Eventually he would abandon science and take up holy orders.

Founder of Modern Geology: James Hutton

The greatest achievement of eighteenth-century geologist James Hutton was to prove that the earth was much older than the 6,000 years claimed by biblical scholars. His view would be highly influential in the formation of Charles Darwin's theory of evolution (*see* p. 147). However, Hutton was unable to provide an exact age for the earth because this required knowledge of the rate of decay of naturally occurring radioactive elements, and radioactivity was unknown at this time.

Hutton first presented his ideas to the Royal Society of Edinburgh during the 1780s and finally published his great work, *Theory of the Earth*, in 1795, two years before he died. Prior to this time there had been some interest in earth sciences, but geology as an individual branch of science was barely recognized.

He claimed that the earth was experiencing a continual pattern of self-restoration and proposed a geological cycle, where the erosion of the land was followed by the deposition of the eroded matter on the sea floor. These deposited particles would consolidate into sedimentary rock, which would rise to form new land and then be eroded again, with the process repeating itself over and over.

In 1787 Hutton observed evidence of this process in the sedimentary rock at Inchbonny, Jedburgh, in what is now known as 'Hutton's unconformity'. The following year he saw similar evidence, also in the Scottish Borders, at Siccar Point, Berwickshire.

From extensive field research on various rock formations, he concluded that the geological cycle was an extremely slow process, since it must have been repeated an indeterminate number of times in the past. Equally, he could find no evidence to suggest that this cyclical process would ever cease and so he assumed it would continue indefinitely.

Hutton's idea that uniform processes acting over immensely long periods of time, and which are responsible for today's features of the earth's crust, will continue into the future and will explain all geological change has come to be known as uniformitarianism, a fundamental concept of geology.

James Hutton (1726–97)

James Hutton's father was an Edinburgh merchant who died when Hutton was very young. After a brief spell as a lawyer's apprentice, the young Hutton decided to follow his interest in chemistry, which led to a course in medicine at the University of Edinburgh, and research in France and Holland.

However, he abandoned medicine and returned to the family farm, raising cash with a friend by manufacturing and selling sal ammoniac, a product made from coal soot that was popular at the time for adding crisp texture to bakery items. He raised enough money for several geology trips round England, France, Belgium and Holland, then settled in Edinburgh in the 1760s, just as the Scottish Enlightenment was flourishing. He moved in social circles that included economist Adam Smith (1723–90), philosopher David Hume (1711–76), chemist Joseph Black (1728–99) and Hutton's future biographer, the scientist John Playfair (1748–1819). Hutton, Smith and Black together founded the Oyster Club, which met weekly to discuss scientific matters. Hutton was also a founding member of the Royal Society of Edinburgh in 1783.

Stimulated by this environment and inspired by the striking physical phenomena he witnessed in and around Edinburgh, Hutton began work on his *Theory of the Earth*.

The Science of Meteorology: John Dalton

In the eighteenth century the patterns in the weather were often still explained by old mythological theories. Weather-watchers or meteorologists who studied the atmosphere were usually amateurs with little understanding of the scientific phenomena that regulated weather, and no systematic approach. The English chemist and physicist John Dalton (*see* p. 118) was largely responsible for changing this attitude and for forging meteorology into a serious scientific activity.

At the age of twenty-one Dalton began to keep a record of the weather he observed and experienced, a habit he continued all his life, but he went further than most weather-watchers by attempting to understand and explain the changes.

In 1793 he published his records of wind velocity and barometric pressure in *Meteorological Observations and Essays*, which also attempted to explain some weather phenomena by discussing the different reactions of gases in the atmosphere. This held the seeds of ideas on atomic theory that would make him a chemical pioneer.

In later meteorological studies on the composition of the atmosphere, Dalton deduced that water, after evaporating, remains in the air as an independent gas. Addressing the puzzle of how air and water are simultaneously able to occupy the same space led him to another chemical breakthrough: atomic weights.

Dalton inadvertently also became an expert on the height of the mountains of the Lake District in north-west England, since in those days the only way to examine temperature and atmospheric pressure at altitude was to climb, estimating altitude with a barometer. Nowadays weather balloons, drones and aeroplanes mean that meteorologists don't have to be so muscular.

Investigating the Palaeozoic Era: Roderick Impey Murchison

A few years after James Hutton's revelations about the age of the earth (*see* p. 201), another Scottish geologist, Sir

Roderick Impey Murchison (1792–1871), attracted fame for his intrepid geological field expeditions and his discovery of the Silurian system, or geological period.

Murchison came from an ancient Scottish Highland landowning family that relocated to England when he was four, after the death of his father. The young Murchison went to military college and served briefly in the Peninsular War, after which he married Charlotte Hugonin, who proved to be a source of great encouragement and inspiration throughout his career.

The couple travelled round Europe and then moved to London, where Murchison attended lectures at the Royal Institution and mixed with famous scientists of the era, including Charles Darwin (*see* p. 147), Charles Lyell (1797–1875) and Adam Sedgwick (1785–1873).

Almost every summer for the next twenty years, the Murchisons embarked upon geological expeditions through Britain, France and the Alps, Charlotte acting as fossil hunter and geological artist.

In 1839 Murchison produced his major work, *The Silurian System*, detailing his research in south Wales on the 'greywackes', or old slate rocks, underlying the 'Old Red Sandstone' sequence of rock and dating from the early part of the Palaeozoic era (Palaeozoic means 'time of ancient life').

Most geologists thought that these slate rocks contained very few fossils but Murchison believed that they could hold the key to discovering the earth's earliest life forms. He named the strata 'Silurian' after the Silures, a tribe that had lived in the region, and found that it marked a major period in the history of life on earth.

Era	Period	Millions of years ago
Cenozoic	Quartenary	5.3
Cenozoic	Tertiary	66.4
Mesozoic	Cretaceous	
Mesozoic	Jurassic	
Mesozoic	Triassic	250
Palaeozoic (Ancient life)	Permian	
Palaeozoic (Ancient life)	Pennsylvanian	320
Palaeozoic (Ancient life)	Mississippian	
Palaeozoic (Ancient life)	Devonian	419
Palaeozoic (Ancient life)	Silurian	
Palaeozoic (Ancient life)	Ordovician	
Palaeozoic (Ancient life)	Cambrian	570
Precambrian		4600

(Phanerozoic Eon spans Cenozoic through Palaeozoic.)

The earth was formed about 4.6 billion years ago. Geologists have subdivided this 4.6 billion years into eons, eras and periods, based upon major changes in the evolution of life on earth according to fossil records. The boundaries between geological eras mark times of mass extinction.

The beginning of the Silurian period has been dated to around 444 million years ago. It has a distinctive fauna (animal life), with many invertebrates but very little in the way of vertebrates or land plants.

Murchison also established, with Sedgwick, the Devonian system in Devon, south-west England, and in the Rhineland,

Germany. Like the Silurian period, it is an interval of the Palaeozoic era; the Devonian period began around 419 million years ago, some 25 million years after the Silurian period. It is sometimes called the 'Old Red Age', after the Old Red Sandstone associated with this time, or the 'Age of Fishes', as it is famous for the thousands of species of fish that developed in Devonian seas, indicated by fish fossils found in the Old Red Sandstone. It was a time, too, when fish first evolved legs and started to walk on land, and the land became covered in forests.

The discovery came out of a controversy. Murchison argued, against geologist Henry De la Beche (1796–1855), that there couldn't be coal below the Silurian system, as the strata beneath must be older than Silurian and coal was associated with younger rocks. Murchison was proved right: the rocks in question were not pre-Silurian but of the newer Devonian system.

Following his famous expedition to Russia in 1840–1, accompanied by Édouard de Verneuil (1805–73) and Count Alexander von Keyserling (1815–91), Murchison defined the Permian system, named after the strata of the Perm region (near the Ural mountains), which has now been dated to between 250 and 290 million years ago.

Ammonites: Lewis Hunton

Lewis Hunton (1814–38) was born and raised on the rugged north-east coast of England. He contributed to the development of the idea that rock layers, or series of strata laid down in chronological order (geological succession), can be subdivided and correlated according to age by analysing the fossils embedded in each layer. This analysis, now known as biostratigraphy, became a fundamental aspect of modern geology.

Hunton's father worked at Loftus Alum Works, a prolific producer of alum made from shale quarried on the high coastal sea cliffs and used to set dye in the textile industry. Undoubtedly these surroundings influenced the young Hunton, who studied geology and fossil zoology in London and met eminent scientists, including Charles Lyell (1797–1875).

Fieldwork for Hunton's first paper, presented to the Geological Society of London in 1836, was carried out across north-east Yorkshire. It provided evidence for his idea that certain fossil species, particularly ammonites (the fossil remains of extinct molluscs), which are abundant in the district's Lower Jurassic rocks, occupied only limited vertical sections of the cliff rock, sometimes only a few centimetres thick, which led Hunton to conclude: 'Ammonites afford the most beautiful illustration of the subdivision of strata, for they appear to have been the least able, of all the *Lias genera*, to conform to a change of external consequences.'

Hunton also excavated a large 5-metre (16-foot) Jurassic-era marine reptile, or ichthyosaur, which can still be seen in Whitby Museum.

Unfortunately Hunton's promising scientific career was cut short when he contracted tuberculosis and died, aged twenty-three.

Mineral Classification: James Dwight Dana

A geologist, mineralogist and zoologist, the American James Dwight Dana played a vital role in the important United States Exploring Expedition of 1838–42. During this geological exploration of the South Pacific he collected an extraordinary range of information on mountain building, volcanic islands, corals and crustacea.

His observations supported Charles Darwin's theory that atoll reefs were the result of coral growth in shallow waters after the subsidence of oceanic islands. Their views opposed other naturalists' ideas that reefs grew where submarine mountains had risen due to accruing plankton debris, leading to a lively scientific debate. It was not until 1951, with evidence from ocean-floor drilling, that Dana and Darwin's theory was proved to be correct.

Dana's reports on the expedition's findings helped raise the USA's status in the world of science, and the samples and collections that resulted from it formed the country's first national museum.

But his major contribution was to provide a classification system for minerals. His method was similar to that used by the Swedish botanist Carolus Linnaeus (*see* p. 142) in his classification for plants and organisms, when he organized them by genus and species. Dana adapted this to produce a revolutionary classification for minerals in which they were organized according to their chemical composition, for example silicates, sulphates or oxides, and then by the structure of the mineral. Dana had four layers in his hierarchy: class, mainly based on composition; type, usually centred on atomic characteristics; group, based on structure; and finally the individual mineral species.

His classification system was a great leap forward for mineralogy, and is universally accepted today. It is adaptable enough that the ongoing discoveries of new minerals can be included within the system, simply slotting into their relevant class and type.

James Dwight Dana (1813–95)

James Dwight Dana entered Yale College (now University) in 1830, where one of his teachers was the prominent mineralogist Benjamin Silliman (1779–1864), founder of the *American Journal of Science*. Dana would later marry Silliman's daughter Henrietta.

After graduating, his first important post was as a mathematics teacher on a US Navy ship that sailed to the Mediterranean. There he observed the volcano Mount Vesuvius erupting, which enabled him to draw comparisons with volcanic activity he observed later in the South Pacific.

In 1836 Dana returned to Yale as Silliman's assistant. He published his mineral classification in *A System of Mineralogy* when he was only twenty-four, two years before embarking upon the United States Exploring Expedition. He then spent ten years writing up his findings, and published his *Manual of Mineralogy*, still an important reference book, in 1848.

Deeply religious, Dana at first resisted Darwin's theory of evolution but then came to accept that it was part of divine intention.

How Landscapes are Formed: William Morris Davis

Geomorphology, or the study of landforms, originated with this American geographer, geologist and meteorologist.

Born to a Philadelphian Quaker family, William Morris Davis (1850–1934) graduated from Harvard University in 1870 with a master's degree in engineering. At this time, very little was understood about how the landscape evolved and how its characteristic appearance developed. With his description of the 'cycle of erosion', Davis was to change this, and with his research and advocacy, he went on to help establish geography as a profession in its own right.

Davis' predecessors believed that the shape of a landform was determined purely by its structure or was created by the biblical flood. Davis, influenced by Charles Darwin's theory of evolution, proposed a system for the development or evolution of landforms that was vaguely similar to that of Darwin. In an article 'The Rivers and Valleys of Pennsylvania', published in *National Geographic* in 1889, he theorized that landscape goes through a long, slow cycle in which first mountains are formed from uplifted land, then over time they are eroded to create V-shaped valleys. As the land further evolves the valleys become wider and rounded hills are formed.

Youth: A young plain is cut with deep V-shaped valleys.

Maturity: The landscape has high slopes and maximum relief, with floodplains.

Old age: Erosion has produced broad valleys that have flattened the remaining hills.

William Morris Davis' cyclical stages of landscape erosion.

Davis described three variables in the evolution of a landscape: structure (shape of the rock and its resistance to erosion and weathering); process (actions such as weathering, erosion, deposition by water); stage (youth, maturity and old age). This last indicated how long the erosion process had been continuing.

Although now considered generally simplistic, Davis' ideas of evolution of the landscape started a new era in the understanding of landforms.

The Pioneering Female Geologist: Florence Bascom

As the first professional woman geologist in the United States, Florence Bascom was a true pioneer for women, both in science and in academia. She had two major achievements: she was an authority on the rocks formed by crystallization in the Piedmont region of Pennsylvania, USA, and she was an inspiration for a new generation of female geologists who came to study under her and who followed in her path.

Florence Bascom was fortunate that both her parents supported women's rights, but after gaining her master's degree in geology from the University of Wisconsin in 1887 she encountered an instructor who was opposed to co-education. She left the university, but opportunities for women were gradually opening up in the USA and she soon went to Johns Hopkins University in Baltimore, Maryland, to study petrology, the science of how present-day rocks are formed.

In 1896 Bascom was appointed geological assistant for the United States Geological Survey, the first woman to have such a role, and was assigned to the mid-Atlantic Piedmont region in Maryland, Pennsylvania, and parts of Delaware and New Jersey. She became an expert in this region, specializing in petrology. During the summer months she mapped rock formations, taking thin sections of rock, and during the winters she analysed these microscopic slides. She learned about the complex, highly metamorphosed crystalline rocks of the area, and her results were published in many folios and bulletins of the US Geological Survey. She was promoted to full geologist for the Survey in 1909.

Her publications brought her widespread recognition and respect, and her work mapping the rock formations in the region became the basis for many future studies.

Florence Bascom (1862–1945)

Florence Bascom forged a path for women both in geology and in higher education. In 1898 she was appointed reader at Bryn Mawr College, Pennsylvania, and became a full professor at the college in 1906, establishing an international reputation for the institute. She was a mentor to an entire generation of young women geologists, three of whom went on to join the US Geological Survey in her footsteps.

Bascom's path was not always easy. While studying for her doctorate at Johns Hopkins University she had to sit behind a screen so that the male students did not know they were studying with a woman. Even on graduation her degree had to be granted 'by special dispensation'. Johns Hopkins only officially admitted women in 1907.

One advantage Bascom did enjoy was that since she was not enrolled as a regular student, she did not have to pay tuition, except laboratory fees.

Continental Drift: Alfred Wegener

The German geophysicist and meteorologist Alfred Wegener was ahead of his time when he proposed the theory of continental drift.

Wegener's first field of study was astronomy, but from 1906 onwards he took part in several expeditions to Greenland to study climatology, and became increasingly interested in palaeoclimatology. He first considered the idea of continental drift in 1910 when he observed the correspondence in shape between the coastlines of the countries on either side of the Atlantic Ocean, in particular South America and Africa. He proposed that originally the earth had one single landmass, a 'supercontinent' that he named Pangaea, then, about 250 million years ago, in the Late Palaeozoic era, this broke apart. Slowly, over time the portions drifted away from each other, which he called continental displacement.

Other scientists had considered whether the American and African continents had once been joined, but had thought that parts of the supercontinent had subsided to form the Atlantic and Indian oceans. Another theory for the fact that many fossils, animals and plants on the two continents are so similar was that there had once been a land bridge linking Brazil and Africa.

Wegener's continental drift satisfactorily explained the similarities in land shape and creatures, but it was controversial since he could not adequately provide a mechanism for how it happened. As a result, at an international conference of geologists in 1928, the theory was opposed in a formal vote, and dropped out of fashion until the 1950s. Then, the emerging field of palaeomagnetism – the study of the earth's changing magnetic fields – and the later development of plate tectonics indicated that, as Wegener had claimed, 'the continents must have shifted'.

Alfred Wegener (1880–1930)

Born in Berlin, Alfred Wegener earned his PhD in that city and in 1905 went to work at the nearby Royal Prussian Aeronautical Observatory, where he first used kites and balloons to study the upper atmosphere. The following year he and his brother Kurt won an international hot-air balloon contest by staying aloft for a world record of fifty-two hours.

On his expeditions to Greenland, Wegener again used balloons to study the climate. During the First World War he served as a junior officer but had long periods of sick leave because he was wounded twice. Afterwards he taught meteorology at Marburg and Hamburg, and he became professor of meteorology and geophysics at the University of Graz in 1924.

In 1930, on his fiftieth birthday and during his fourth Greenland expedition, Wegener never returned from a routine supply check. His frozen body was later found and his death was attributed to heart failure.

New Technology and Climate Change

Satellite imaging and recordings from pilotless aircraft are having a major impact on both geology and meteorology, giving both disciplines perspectives that only a few decades ago scientists could never have imagined.

Geological features can now often be seen in their entirety, even fault lines and the changes due to plate tectonics, helping seismologists predict earthquakes with greater accuracy. Satellite imagery is also a cost-effective method of oil and gas exploration.

Using earth-orbiting satellites, meteorologists can see weather patterns developing over wider areas and collect information about our planet and its climate on a global scale. Over many years, the data can reveal signs of climate change.

Most climate scientists believe that the current global-warming trend, which is accelerating at an unprecedented rate, is likely to be human-induced, caused by activities such as deforestation, burning fossil fuels and the use of fertilizers. The increased levels of carbon dioxide and other 'greenhouse gases' in our atmosphere act as a type of thermal blanket, absorbing heat radiating from the earth's surface, and sending it back down to warm the earth. Carbon-dioxide levels have increased by a third since the Industrial Revolution.

Evidence of climate change includes rising sea-levels, global temperature increases, warming oceans, shrinking ice sheets and extreme weather.

The general consensus among climate scientists is that climate change is occurring, and the geological evidence shows that climate change can happen relatively quickly, within a century or even a few decades.

Index

Illustrations in italics

A
Adler, Alfred 193
aether 16, 112
Agricola, Georgius 198–9
alchemy 71, 112–13, 114, 172
Alcmaeon of Croton 167
Alexander the Great 18
Alfonsine Tables 27
algebra 62, 63, 64, 69, 72
algorithms 78
alleles 153–4
Altman, Sidney 161–2
amino acids 128, 131
ammonites 208
amoebas 146
Ampère, André-Marie 89
analytical geometry 63, 64
anthrax 180–1
anti-Semitism 29, 99, 103, 106, 151, 193
antibiotics 186–9
Archimedes 54–6
argon 129
Aristotle 15–17, 82, 112, 118, 138–9, 152, 196
Arithmetica (Diophantus) 64, 77
artificial intelligence 74–5
Aryabhata 58–9
aseptic surgery 182–3
astrolabes 27, 28–9
astrological medicine 33
atomic bombs 103, 106, 108–9, 132
atoms
    discoveries in chemistry 118–19, 124–5, 130
    discoveries in physics 87, 95, 98, 101–2, 104, 105, 109
Avicenna 173–5
Avogadro, Amedeo 87–8
axis, earth's 18–19, 25
Azarquiel 26–7

B
Babbage, Charles 78
Babylonian civilization 13, 20, 112
bacilli 181
Bacon, Francis 114
Bacon, Roger 10
bacteria 141, 150–1, 160
al-Baitar, Ibn 175–6
Bascom, Florence 213–15
al-Battani 24–5
batteries 86–7, 117
Becquerel, Henri 95, 97
Bell, Susan Jocelyn 49
Berg, Paul 159–61
Berners-Lee, Tim 78–9
Berzelius, Jons Jacob 120, 122
Betelgeuse 46
'Big Bang' theory 42, 45
binary system 68
biostratigraphy 208
birth control *see* contraception
black holes 48, 48–9, 50–1
blackbodies 95, 103–4
bloodletting as cure 170
Bohr, Aage 103
Bohr, Niels 102–3
Boltzmann, Ludwig 127
bonding, chemical 130–1
Book of Simple (Herbal) Remedies (al-Baitar) 176
Bose-Einstein statistics 104
Bose, Jagadis Chandra 99–101
Bose, Satyendra Nath 103–4
bosons 104, 110
Boyle, Robert 114–15
Brahe, Tycho 36
brain physiology 183–6
Broglie, Louis de 106
Brownian motion 98
Bruno, Giordano 39
Bunsen, Robert Wilhelm 123–4
Bunsen burner 123

C
Caesar, Julius 15
calculus 68, 69–70
calendars 15, 19–20, 58
cancer 160, 179
Canon of Medicine (Avicenna) 173
Carson, Rachel 162–3
Cartesian coordinates 61–2, *62*
Cassini, Giovanni 40
cathode rays 88, 93
Caton, Richard 183
Cech, Thomas R. 161–2

cells and cell theory 145–7, 179
Celsius, Olof 142
CERN 78, 106, 110
Cesalpino, Andrea 142
Chain, Ernst 187
Chandra (Subrahmanyan Chandrasekhar) 47–9
chaos theory 74
Charcot, Jean-Martin 192
chemical potential 127
chemical weapons 120
China 112, 165, 197–8
chromosomes 146–7, 155–6, 157
chronometers 30–1
classification of plants and animals 137, 138–9, 142–4, *144*
classification of stones and minerals 197, 198, 209–10
climate change 196, 197, 198, 217–18
Cohn, Ferdinand 150–1
Columbus, Christopher 28, 29
compasses, magnetic needle 26
compounds, chemical 111, 115, 117, 121–2, 127–8, 131
computers 58, 66, 68, 74–6, 77, 78–9
constellations 13, 15
continental drift 215–16
contraception 194
Cook, Captain James 31, 143
Copenhagen Interpretation 105
Copernicus, Nicolaus 24, 31–3, 34
Corey, Elias James 132–3
covalent bonding 130
Crick, Francis 157–8
Curie, Marie 95–7
Curie, Pierre 95–6, 97
'cycle of erosion' 211–13

D

da Vinci, Leonardo 82, 170
Dalton, John 87, 118–19, 203–304
Dana, James Dwight 209–10
dark energy 45
dark matter 14, 46–7
Darwin, Charles 138, 139, 147–50, 201, 205, 209, 211
Davis, William Morris 211–13
Davy, Humphry 86, 89–90, 117–18
De la Beche, Henry 207
de Lamarck, Chevalier 152
De Natura Fossilium (Agricola) 198
De Re Metallica (Agricola) 198

decimal places 59
Declaration of Independence 85
Democritus 118
Descartes, René 61–4, 67, 183
Devonian system 206–7
dietary suppliments 190
digital revolution 68
Diophantus 64
Discourse on the Method (Descartes) 63–4
Disquisitiones Arithmeticae (Gauss) 72
dissection of bodies 165–6, 167, 168, 169, 170
DNA (deoxyribonucleic acid) 131, 157–9, *158*, 160–1
Dumas, Jean-Baptiste 121–2

E

earth, age/chronology of 195, 197, 200–2
Eddington, Arthur 41, 42
Edison, Thomas 10–11
Egyptian civilization 13, 57, 165, 166–7, 194
Einstein, Albert 41, 43, 91, 97–9, 103–4, 108
electric motors 89–90
electricity 11, 84–7, 88–9
electrolysis 90, 117
electromagnetism 89, 90, 91–2, 95
electron degeneracy principle 48–9
electrons 88, 101–2, 125, 130, 131
Elements of Geometry, The (Euclid) 53–4
$E=mc^2$ 98
Enceladus' geysers 40
endorphins 186
endospores 150–1, 181
Enigma Code 76
environmental pollution 162–4
enzymes 161–2
equinoxes, precession of the *19*, 19–20, 58
Euclid 53
eugenics 149
evolution, theory of 147–9, 211
expansion of universe 42, 44–5
Explanation of Binary Arithmetic (Leibniz) 68

F

Faraday, Michael 89–90, 123
femtochemistry 134–5
femtoseconds 134
Fermat, Pierre de 62, 64–6, 76–8

INDEX

Fermi, Enrico 109
fermions 104
fertilizers 120
Feynman, Richard 108
Fibonacci 59–60
Fibonacci Sequence 60–1, *61*
First World War 96, 120, 188
Fischer, Emil 127–8
Fleming, Alexander 186–9
Florey, Howard 187
fossil plants 197–8
Franklin, Benjamin 84–5
Franklin, Rosalind 157
free radicals 130
French Revolution 116–17
Freud, Sigmund 192–4

G
Galen 169–71
Galilei, Galileo 10, 15, 37–40
Galton, Francis 149
Galvani, Luigi 86
gametes 153–4
Gan De 14–15
gases, Boyle's law on 114
Gauss, Carl Friedrich 71–3, 88–9
Gay-Lussac, Joseph Louis 119
genetic engineering 159–62
genetics 152–61, *154*, *155*
genomes 159
geocentric theory 16, *17*, *22*, 20–3
geometry 53–4, 57, 62, 63, 65, 71
geomorphology 197, 211–13
Gesner, Konrad von 142
Gibbs, Josiah Willard 92, 126–7, 129
global warming 218
Goethe, Johann Wolfgang von 31–2
golden ratio 54
Golgi, Camillo 184–5
gravity 35, 41, 46–7, 48, 69, 83
Greece, ancient 10, 87, 118, 165, 166–8, 194
see also Aristotle
Guericke, Otto von 114

H
Haber, Fritz 120
Hahn, Otto 109
Halley, Edmond 41
Harrison, John 30–1
Hawking Radiation 51
Hawking, Stephen 50–1

al-Haythem, Ibn 10
healers, Egyptian 166–7
Heisenberg, Werner 104–6, 107
heliocentric theory 31–2, 33–6, 39
heliotrope 88
herbalists 175–7
Herodutus 166
Hertwig, Oscar 147
Hertz, Heinrich 92–3
Higgs boson particle 110
Higgs, Peter 109–10
Hipparchus of Nicaea 15, 18–19, 23, 27
Hippocrates 167–8, 169, 170
Historia Naturalis (Pliny the Elder) 198
Homer 166
Hooke, Robert 141, 200
Hoover, Herbert and Lou 198
Hubble, Edwin 42, 43–5
  1936 classification of galaxies 43
Human Genome Project (HGP) 159
Humboldt, Alexander von 88
humours 168, 169, 172–3
Hunton, Lewis 207–8
Hutton, James 195, 201–3
Huygens, Christiaan 40
hydrometers, invention of 112
Hypatia 112–13
Hypertext Markup Language (HTML) 79
Hypertext Transfer Protocol (HTTP) 78–9

I
immunology 177–9
India 118, 165
infinitesimal calculus 63
Internet 78–9
ionic bonding 131
isomers 120

J
Jansenism 67
Jenner, Edward 177–9
Jesty, Benjamin 178–9
Jung, Carl 193
Jupiter 13, 14, 15, 32, 38

K
Kepler, Johannes 34–7
  Second Law 35
Kirchoff, Gustav 124
Knidian medicine 168
Koan medicine 168
Koch, Robert 150, 180–2

L

landscapes, formation of 211–13, *212*
Large Hadron Collider 110
lasers 134
Lavoisier, Antoine-Laurent 115–17
Leavitt, Henrietta 42–3
Leeuwenhoek, Antoine van 139–42, 150, 177
Leibniz, Gottfried Wilhelm von 68, 70
Lemaître, Georges 42
Leucippus 118
Lewis, Gilbert N. 129–30
Liber abaci (Fibonacci) 59, 60
Liebig, Justus von 119–20, 121, 122
light and gravity 42
light as electromagnetic radiation 91–2
light, quantum theory of 97
light, speed of 98
Lindemann, Ferdinand von 58
Linnaeus, Carolus 137, 142–4, 209
lobotomies 184
longitudinal problem 30–1
Lorenz, Edward 74
Lovelace, Ada 78
lunar eclipses 28
Lyell, Charles 149, 205

M

magnetism 88–9, 90
magnetoelectric induction 90
magnetometer 88
Manhattan Project 103, 108–9
Manual of Mineralogy (Dana) 211
Marconi, Guglielmo 93
Mars 13, 16, 32, 34
Mary the Jewess 113
matrix mechanics 105
Maxwell, James Clerk 91–2, 127
McClintock, Barbara 156
medical ethics 172
meiosis 147, *147*, 153, 155
Meitner, Lise 109
Mendel, Gregor 152–5
Mendeleev, Dmitri 124–6
mental illness 184
Mercury (planet) 13, 16, 32
metabolism 137
metallurgy 198
meteorology 203–4
Meteorology (Aristotle) 196
Meterological Observations and Essays (Dalton) 204
method of exhaustion 55
microbes 181
Micrographia (Hooke) 141
microorganisms 137, 150, 174, 177, 182
microscopes 137, 139–42, 151
microwaves 99
Miescher, Johannes Friedrich 157
mining and minerals 198–9, 209–11
mitosis 146, *146*
modularity theorem 77
molecular diseases 189–90
molecules 87–8, 112, 119–20, 127, 131, 134–5
moon 16, 24, 32, 38
Morgan, Thomas Hunt 155–6
motion, laws of 83–4
motors, electric 89–90
Murchison, Roderick Impey 204–7

N

n-body problem 74–5
natural selection 148
navigation, maritime 26, 27–30
negative numbers 59, 60
nervous system 184–5
neurones 183, 185, *185*
neuroscience 183
neutrinos 109
neutron stars 46, 48, 50
neutrons 102, 109
Newlands, John 124
Newton, Isaac 10, 35, 40–1, 63, 68, 69–71, 73, 83–4, 110
Nobel Prizes 47, 97, 103, 132
noble gases 128–9
North Star 25
nuclear fission 109
nuclear weapons 108, 132
nuclei 102, 109, 130, 145, 157
nucleic acids 161
nucleotide pairs 158–9
number theory 64–5, 72

O

On Stones (Theophrastus) 197
On the Equilibrium of Heterogeneous Substances (Gibbs) 126–7
oncology 179
Oppenheimer, J. Robert 108
optics 83
Ørsted, Hans Christian 89
Orta, Garcia de 176

orthomolecular medicine 190
Oscar II of Sweden, King 73
oxygen and combustion 115–16
Oyster Club 203

P

palaeoclimatology 197, 215
palaeomagnetism 216
Palaeozoic era 205, 207
Pangaea 215
Pascal, Blaise 64, 65–7
   theorem of projective geometry 66
Pascaline calculating machine 66, 78
Pascal's triangle 65
Pasteur, Louis 181–3
Pauli exclusion principle 48–9
Pauling, Linus 130–2, 189–90
Pavlov, Petrovich 184
penicillin 186–9
Penrose, Roger 50
Pensées (B. Pascal) 67
periodic table 111, 124–5, 129
Permian system 207
Pert, Candace 186
pesticides 162–3, 164
Philip of Macedon 17–18
photons 97, 104
photosynthesis 120
pi 57, 58
Pincus, Gregory Goodwin 194
Planck, Max 94–5, 103–4
planets 13–16, 18–25, 31–2, 33, 34–6, 37–8
plant biology 144, 152–3, 156
plant chemistry 120
plant medicine 175–7
plant physiology 100
plate tectonics 216
Plato 17, 139
Platonic solids 54
Pliny the Elder 198
Poincare, Henri 73–5
Polaris 25
pole star 25–6, *26*
polio 190–2
poverty and disease 180
Priestley, Joseph 115, 116
prime meridian 31
Principia Mathematica (Newton) 41, 69, 71, 83
Principles of Geology (Lyell) 149
probability theory 64, 65, 66

projective geometry 65
protein molecules 131
Protestantism 36
protons 102, 125
psychoanalysis 192–3
Ptolemy 13, 20–3, 32
purines 127–8
pyramids of Giza 57
Pythagoras' theorem 54, 59

Q

quantum mechanics 105, 130
quantum statistics 104
Quantum Theory 94–5, 97
quarks 102
quasars 49

R

radio waves 92–3
radioactivity 95–7
Ramen, C. V. 47
Ramón y Cajal, Santiago 184–5
Ramsay, William 128–9
Rayleigh, Lord 129
relativity, theories of 41, 48, 91, 98
religion 23, 29, 36–7, 39, 67, 149, 165, 194, 199, 200–1
   see also anti-Semitism
retrosynthetic analysis 133
Rhazes 171–3
RNA (ribonucleic acid) 161–2
Roman civilization 13, 169–70
Röntgen, Wilhem 93–4
Roosevelt, Franklin D. 108
Royal Society 71, 141, 142, 178
Royal Society of Edinburgh 201, 203
Rubin, Vera 47
'rules of reasoning' 10
Rutherford, Ernest 96, 101–2

S

Sabian Tables 24
Sabin, Albert 191
Salk, Jonas 190–2
satellite imaging 217
Saturn 13, 16, 32, 39–40
Saturn's rings 39–40
Sceptical Chymist, The (Boyle) 114
Schleiden, Matthias 145
Schrödinger, Erwin 106–8
Schrödinger's Cat 107
Schur, Max 194

Schwann, Theodor 145
Second World War 76, 103, 106, 108
Sedgwick, Adam 205, 206
Shen Kuo 25–6, 197–8
Shi Shen 14, 15
short radio waves 99
sickle-cell anaemia 189–90
Silent Sprint (Carson) 162, 164
Silurian system 205–6, *206*, 207
Silurian System, The (Murchison) 205
sines, tables of 58
singularities 50–1
smallpox 178–9
Smoluchowski, Marian 98
Socrates 18
Solander, Daniel 143
solar eclipses 25, 41
solar tables 28
Sosigenes of Alexandria 15
space exploration 11
Species Plantarum (Linnaeus) 143
spectroscopy 124
Spirogyra algae 140–1
star catalogues 14–15, 18
statistical mechanics 127
statistics 103–4, 127
Steno, Nicolaus 200–1
subatomic particles 88, 103, 104, 105, 106–7, 110
sun, the 16, 24, 28, 30, 38
 see also heliocentric theory
supernovas 45–6, 48, 50
surgery 165–6, 169–70, 182–3, 184
synthetic chemistry 128, 132–3
Szilard, Leo 108–9

T
Taylor, Richard 78
telescope, invention of the 13–14, 37–8, 39
Thales of Miletus 82
Theophratus 138, 196–7
Theory of the Earth (Hutton) 201, 203
thermodynamic chemistry 92, 126–7, 129–30
Thomson, J. J. 88, 101
three-body problem 73–4
time, measuring 15, 19–20, 24
Toledan Tables 27
trepanation 165
trigonometry 58–9
Turing, Alan 74–6
Turing Test 75, 78

U
uncertainty principle 104–5
unicellular organisms 140–1, 146, 174
uniformitarianism 202
uranium 95–6, 109

V
vaccines 177–9, 181, 190–2
vacuums 66, 114
Venter, Craig 159
Venus 13, 16, 32, 38
Vespucci, Amerigo 30
Viète, François 64
Virchow, Rudolf 179–80
Volta, Alessandro 86–7
voltaic piles 86–7, *87*, 89, 117
Voltaire 68
voltameters 90

W
Wallace, Alfred Russel 148
war machines, Archimedes' 55, 56
Watson, James 157–9
wave mechanics 106–7
Weber, Wilhelm 89
Wegener, Alfred 215–17
Wei Pu 25–6
white dwarf stars 47, 48
Whitson, William 200
Wiles, Andrew 76–8
Wilkins, Maurice 157
Wöhler, Friedrich 119–20
World Wide Web 78–9

X
X-rays 93–4, 96, 131, 157

Y
years, measurement of 15, 19–20, 24

Z
Zacuto, Abraham 27–8, 29
al-Zarqali 26–7
zero, concept of 59–60
Zewail, Ahmed H. 134
Zhang Heng 57
Zu Chongzhi 57–8
Zwicky, Fritz 45–7, 50
zygotes 147